ENGLISH FLOWERS
FROM FOREIGN FIELDS

THE ORIGINS OF ENGLISH WILD FLOWERS

COLONISTS
Plants from agricultural activities, usually no other habitat than arable land.

DENIZENS
Plants introduced by man in farm and garden which escaped from cultivation and became part of the natural vegetation.

ALIENS, CASUALS AND ADVENTIVES
Plants introduced accidentally by man through various methods of transport.

NATIVE SPECIES
No human action
Usually here before man.

ENGLISH FLOWERS FROM FOREIGN FIELDS

WHEN THEY CAME AND WHERE THEY WENT

Elizabeth Adlam

Cappella Archive
Limited Editions

Cappella Archive
Foley Terrace : Great Malvern : England

Printed on Demand : November 2000

Copyright © Elizabeth Adlam 1999
All rights reserved

No part of this publication may be reproduced or stored in a retrieval system, nor transmitted in any form or by any means – digital scanning, mechanical, photocopying, recording, or by any other media – without prior permission of the copyright owner.

British Library Cataloguing-in-Publication Data
A catalogue record for this book is
available from the British Library

ISBN 1–902918–01–0

Typeset in a Cappella realization of a Baskerville of 1769 and printed on Five Seasons paper from John Purcell of London.

CONTENTS

	PROLOGUE	ix
1	WHERE DID IT ALL BEGIN?	1
2	THE GREAT ROCK GARDEN	7
3	TILL, SOW, AND REAP	14
4	VIATORES	23
5	THE DARK BEFORE THE LIGHT	32
6	1066 ETCETERA	37
7	AS WE LIKE IT	49
8	THE GREAT PLANT WRITERS	58
9	THE FLORAL HOLOCAUST	66
10	FOR BETTER, FOR WORSE	76
11	THE CARPET BAGGERS	83
12	ABOUT THE HOUSE	89
13	ON OUR CONSCIENCE	94
14	NEW PLACES TO HIDE	100
15	THE NEXT STEP	106
	EPILOGUE	113
	APPENDICES	114
	PLANT INDEX	118
	FURTHER READING	xi

COLOUR PLATES
photograps taken by the author

1. False Oxslip
2. Wistman's Wood
3. Primroses
4. The Roman Roads
5. An Old Salt Route
6. Hybrid Hollyhock
7. Houseleek
8. Oxford Botanic Gardens
9. Wild Daffodils
10. Feverfew
11. Yellow Flag Iris
12. Wild Arum
13. God's Acre
14. Ragwort
15. Polyanthus x Cowslip
16. A Landscape Mosaic

ILLUSTRATIONS

Wild Flower Origins	ii	The Duke's Tea Plant	63
Sources of the Weather	x	Flax	65
Time Chart	4–5	A Pre-enclosure Landscape	68
Gorse	6	A Post-enclosure Landscape	69
Pollen Proportions	8	Rectangular Fields	72
English Moorlands	13	Greater Plantain	75
Charlock	16	Himalayan Balsam	81
Betony	20	Coltsfoot	82
Borage	22	Hemp	85
Roman Agger	24	Knotgrass	88
Collecting Opium	27	Woodruff	93
Scarlet Pimpernel	31	Corn Marigold	97
Bellflower	41	Stinging Nettle	99
Scabious	45	Corn Cockle	104
Gerard's Herbal	48	Community Forests	105
Vetch	53	Wellingtonia	109
Sainfoin	57	Bluebell	112

PREFACE

Going for walks with my father are amongst my earliest childhood memories. At this distance in time, I cannot decide if it was his way of occupying a restless little girl, or a response to the current petrol rationing. It was probably both. We walked for many hours together, with my little hands clutching a small *Flora* in which all the flowers were grouped according to colour; ideal for a small child. Long before I went to school, I slowly learned their names and copied them into a small exercise book.

My fascination for flowers and their histories grew into a lifetime's interest, and this book is an attempt to tell the story of where they came from and how we have used them through the ages. I hope it will give pleasure to those who read it, and encourage those who become curious to research for themselves the extraordinary world of wild flowers.

Without doubt, the most heartfelt thanks must go to my long- suffering and supportive family. Their tolerance, sometimes very strained, has always been welcome. My appreciation also goes to my many friends, whose unflagging interest and support has been a great boost, especially when the going got rough and the end never seemed to be in sight.

The final transformation into a book from a rough outline of the text, scattered drawings, and sundry photographs, was achieved by my publisher and long-time friend, David Byram-Wigfield, to whom I give my grateful thanks. His tolerance and patience in explaining the subtleties of printing techniques has been enormous. Finally, I must thank my husband, Brian, who prepared all the maps and diagrams in the book, as well as many others which have not been used in the final text. His support in the final stages of preparing this book has been tremendous.

ELIZABETH ADLAM
ALFRICK 1999

DEDICATION

For my late father, who introduced me to the
wonderful world of flowers.

PROLOGUE

If you can look into the seeds of time
William Shakespeare

Out of the mists of time they came, marching over the land like an elite corps of infantry. Highlands and lowlands, valleys and plains, they all surrendered to the invading plant kingdom.

Under their protection the first land animals, the amphibians, fed, grew and roamed the land. The reptiles which followed ate them too, and even some of the large dinosaurs, for not all of them were carnivores.

At the same time, the plants, like the animals, were changing their form. By the time modern man appeared, walking out of Africa into Europe as *Homo sapiens sapiens* some 150,000 years ago, most of the present-day plant species had arrived.

During our early development, we had walked with plants, which fed and protected us. As civilizations rose and fell, they were ever-present. We walk with them now, and they will be present when we are gone.

This is their story; how they have supported us; travelled along with us, and will take us over, if we do not give them the respect they deserve.

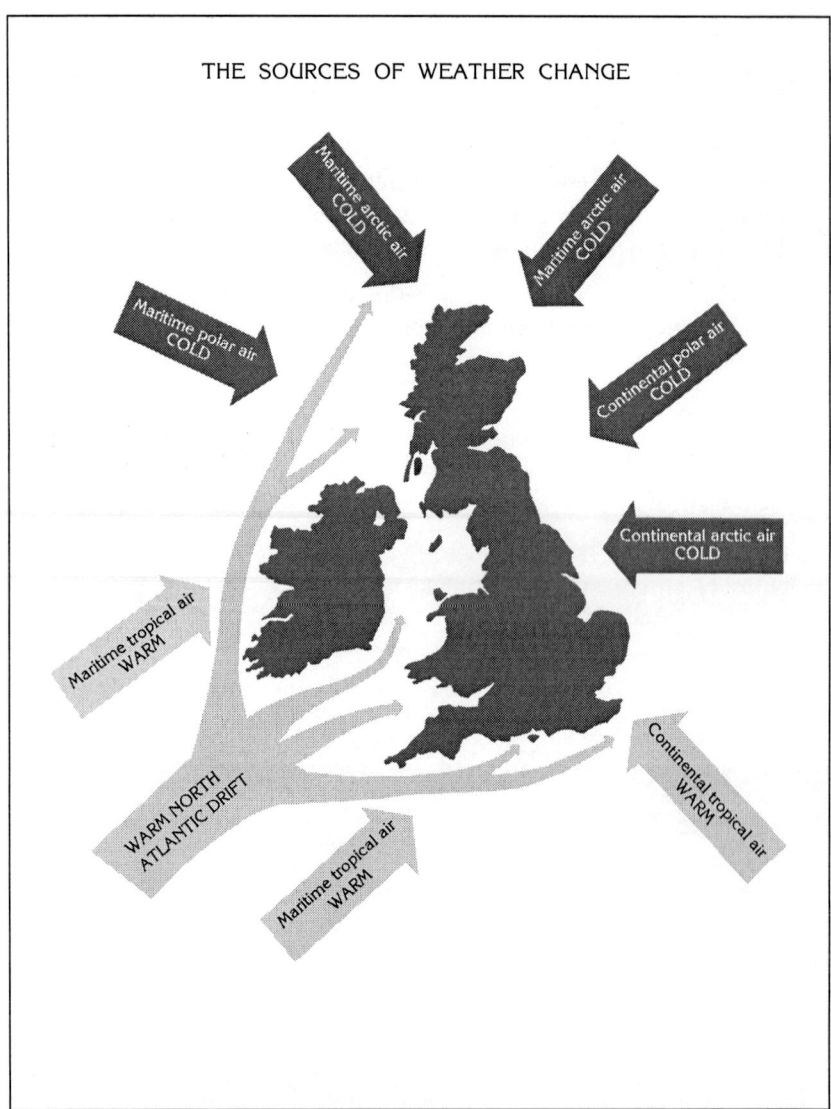

1
WHERE DID IT ALL BEGIN?

God made the country and man made the town.
William Cowper

The United Kingdom today is a largely urbanised society, self-assured in its increasing technological skills. The gap between us and the plant kingdom, with which we evolved, is also increasing as we appear to be losing our past plant connections. The contact is still there though often sub-conscious. Comments such as 'Just going for a walk in the park' and 'Must do half an hour in the garden', show that there is still a bond. It can also be seen every weekend. The big cities empty as their inhabitants make for the peace and quiet of the countryside, and long queues wait to enter the car parks of the numerous garden centres where there are garden plants, pot plants and cut flowers for sale.

The kingdom of plants with its recurring annual rhythms has always fascinated man. Its beauty was here before he evolved and will be here when he has gone. It has associations with life itself, so it is not surprising that the world's great religions, Christianity, Islam, Hinduism, and Buddhism, are inseparable from plants. 'Say it with flowers' is a modern advertising slogan, but man has been doing it since he first had time to appreciate his surroundings. The flowers in the Egyptian tombs and the Neanderthal burial sites are two well-known, historical examples of floral tributes.

In the beginning man fed on fruits and seeds, followed by the grains such as wheat, barley, and rye, which also fed his domesticated animals. Trees were cut down and burned to keep him warm and to provide fuel for cooking. They were made into houses as well as bows and arrows for defence and to kill animals for food. Man made clothing from cotton, flax, hemp and stinging nettles. Plant extracts were medicines for illnesss and soothing the mind into sleep. Eventually the plants became a potent power over our minds and bodies, for they became credited with powers of good and evil. The good powers are often seen in their names such as 'eggs and bacon' and 'town hall clock'. A plant often associated with both good and evil is mistletoe due to its growth on a tree with no apparent roots.

It is possible to view some of England's social and economic history through the plants which are grown today for food. A good place to begin is the traditional Christmas meal. Before the festive meal a drink is served. Pimm's is popular, especially when garnished with borage which arrived here with Iron Age Man and was used as a fever cure. On the menu for the main course are potatoes, parsnips, peas, swedes, and turkey, served with a variety of herb stuffings such as sage, parsley, rosemary, thyme and sweet chestnut. Not many items on this list are native English food. The potatoes—and the turkey—are from the New World, whilst parsnip, sage, rosemary, sweet chestnut, and culinary thyme are all native Mediterranean plants and Roman introductions. On the dessert menu there is fresh fruit salad containing cultivated cherries, apples and pears—which have non-native Mediterranean ancestors, usually Roman in origin—native hybridised raspberries, strawberries with red or black currants. Cream if necessary would come from locally raised European breeds of cow.

This example shows that the plants which are grown here for food have several non-native origins. The sweet chestnut is one of the many introduced food plants which left the safety of the garden to become naturalized in the wild. The frontispiece of the Origins of English Wild Flowers shows the usual classification of the English flora by origin, clearly indicating where the details of our past history can be found. Plants have arrived here by many different means becoming naturalized with our native flora. As a result it becomes clear that the history of our flora is a multi-coloured strand with numerous threads.

When artificial hybrids from native plants are included in this history, the effect is even more intriguing. Where would gardeners be without those hybrid wallflowers, which were bred from the alien, casual, adventive plant of Mediterranean origin, which arrived here from the Caen region of France on building stone? Then there is the native Cheddar pink, which was crossed with the Roman carnation, giving a huge range of hybrid, brightly coloured, frilly-petalled flowers which fill every florists' shop. Natural hybrids also appear, for example the crosses of primroses and cowslips as well as natural 'sports' or 'rogues', recorded since Elizabethan times. These have become the basis of the huge range of the modern brightly coloured polyanthus.

A multi-coloured strand like this will only hold together if it has a good glue. Here the adhesive is the climate of north-western Europe. It is a soft climate—although at times it may not feel like it—producing the best environment for the growth of temperate zone plants. There is rain throughout the growing season and almost twenty hours of daylight in the summer months. Although England is on the same latitude as Moscow and Winnipeg, there is no long, cold winter. The westerly air streams bringing in the warm oceanic air, combined with the gentle heating effect of the North Atlantic Drift, contribute to the mild climate and ice-free coastal waters of western Europe.

Even with this relatively stable climate there have been some fluctuations. Since the end of the last Great Ice Age our climate has oscillated around an average not unlike the average for the twentieth century. The Bronze Age was a time of warm dry weather which was followed by a period of damp, cool weather prior to the Roman invasion. It was hotter and drier here in Roman times allowing the extensive cultivation of vines at the peak in the fourth century AD. Probably this warm period, in England as well as Europe as a whole, started the downfall of the Roman Empire. Nomadic people in drought areas of of eastern Europe, migrated west searching for food. These invading hordes produced the Roman withdrawal of manpower from their northern and eastern provinces, since the troops were needed to defend Rome. Eventually the whole civilisation fell under this pressure as well as the internal wrangling and power struggles.

There were little Ice Ages in the eighteenth and nineteenth centuries, which produced Frost Fairs on frozen rivers such as the Thames and the Dickensian snowy Christmases. This century the cold winters of 1962–3 and 1978–9 were nowhere near as severe as the little Ice Ages, nor were the hot summers of 1921, 1976, and 1995 comparable to the fourth century warm period. As far as can be estimated, the general overall effect of these climatic variations was not devastating. Fossil and written evidence would suggest that few if any plants have been recorded as completely lost, though the distribution of many species at the edges of their range was frequently reduced.

It is possible to determine whether a plant existed on a site even if there is no written evidence. Fortunately many soils, especially peat ones, have a low level of oxygen which means that the rate of plant decomposition is very slow and the preservation of readily identifiable plant material. Microscopic analysis of these soils produces a range of macro and micro fossils. The macro fossils are fragments of plants such as seeds, roots, and twigs approx. 0.5–0.1mm. Pollen and spores form micro fossils of 10—150 μm and are less fragmentary.

Pollen is extremely reliable evidence for a plant's existence, for its shape and markings are as unique and individual as a set of fingerprints, allowing accurate identification. Due to its very light weight pollen is very easily blown about, so occurrence is not always an indication of plant distribution.

Soils from airless or waterlogged bogs, agricultural land, around old dwelling sites, waste pits, and latrines can all be analysed for their plant fossils. In this way it is possible to determine what was growing in that region, what was eaten, and what was thrown away. Excavations at Windsor Castle after the recent fire revealed the diet of the thirteenth and fourteenth century courts. The stomach contents of people and animals preserved in ice and bogs can be analysed to see what had been eaten before death. However, plants which produce little pollen or live in dry places may have a poor or non-existant pollen record.

The natural English vegetation is deciduous, broadleaved woodland with predominantly seed-bearing herbs, trees, shrubs, and herbs. The spore-producing plants such as ferns, mosses, liverworts and lichens are fewer in number and less dominant. The evolution of seed-bearing plants started at the beginning of the Tertiary period some 65 million years Before Present; known as years BP. The sub-tropical climate had increased the rate of development of the sub-tropical flora producing a large and varied vegetation. However before the following Pleistocene or Great Ice Age—750 thousand BP—the flora had begun to change to a more temperate one, as the earth cooled down. Most of the tropical species, mainly trees, were lost, never to return by natural means.

During the long cold Great Ice Age from 750 thousand BP to 12 thousand BP, it was the temperate zone plants which wintered in Europe as refugees, returning as the ice finally retreated. The tropical and sub-tropical plants found safe refuge on the land which would become North America and Asia. They were unable to return when the ice retreated, due to the physical barriers of new oceans and

Late Palaeolithic	10	9	8	7	6	5	4	3	2	1	Present Day
PRE-BOREAL		BOREAL			ATLANTIC			SUB-BOREAL		SUB-ATLANTIC	
Warmer Drier		Warm Dry			Warm Wet			Warmer Drier		BC-AD Cooler Wetter	
				Mesolithic		Neolithic		Bronze Age	Iron Age	Romano–British	

Thousands of years BP

Climatic Changes Before Present

Where did it all begin?

mountains. In this tropical flora were species such as *Sequoia*, *Tsuga*, *Taxodium*, *Pterocarya*, *Carya*, and *Nyssa*. Curiously these trees, according to the fossil record were only 5% of the late Tertiary flora. The dominant plants were those which are present here today and are known as native or endemic species. These include Scots pine, yew, hornbeam, hazel, birch, alder, beech, oak, elm, mountain ash or rowan, sloe, hawthorn, field maple, and dogwood. It becomes increasingly probable that our flora was influenced more by the change from a tropical to a temperate climate over all north-west Europe. This left most of Great Britain much as it is today in terms of its climate and range of vegetation. The effect of the Pleistocene or the Great Ice Age was to interrupt the distribution of plants driving them south to the European refuges, from where they returned once the ice finally retreated.

Curiously, with the return to a temperate climate some of the exotic or non-native species have also returned, not in the wild but in our gardens. They survived the Great Ice Age in refuges on other continents such as North America. *Sequoia* or redwood, the tallest tree in the world, honours a half-breed Cherokee chief Sequoyah. The native habitat is the Pacific Coastal region of the USA from San Francisco to Oregon. It was first introduced into the United Kingdom in 1843 via Russia, an unusually long return route! It is found in many public parks and private gardens, especially those of the Victorian era. Its sudden appearance in the middle of many present day housing estates, indicates that the land was once a park or garden, planted by our tree-loving Victorian ancestors.

Another popular Victorian tree *Tsuga* or western hemlock is also native to the Pacific coast of the USA and was first introduced back here by John Jeffrey in 1853. The swamp cypress, irreverently known as 'knobbly knees' or *Taxodium*, is a member of the same family as the redwood and western hemlock but it is deciduous and comes from the

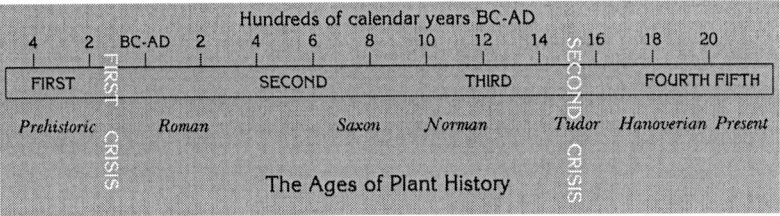

The Ages of Plant History

eastern coast of the USA. It is successfully planted in south-eastern England, especially in the London suburbs, but is rarer further north, for it needs warmer summers for proper growth to be maintained.

Pterocarya – wingnut – and *Carya* – hickory – are the less well-known members of the walnut family, which were first introduced here in 1782 from the USA via the Caucasus. This was another circuitous route for both these trees from the eastern USA, from where several of our modern garden trees have originated. Today these two trees are rare garden exotics in the south of England.

In effect the wheel has turned the full circle allowing the return of some of the natives. This was driven by man's insatiable urge to collect and his enormous desire to travel. It has led to the collection of plants from medieval times to the present day, from places as far apart as China, southern Europe and North America. The range of exotics, planted from Scotland in the north to the Home Counties in the south, shows the present day variations in the British climate, perhaps indicating what it was like here immediately before the Great Ice Age.

Gorse

1. *False oxslip, a natural hybrid*

2. *The ancient forest of Wistman's Wood*

3. *Primroses*

4. *The Roman Roads*

2
THE GREAT ROCK GARDEN

Tomorrow to fresh woods and pastures new.
John Milton

At the end of the Great Ice Age, around 14,000 BP, the ice began to melt and started its last retreat north, revealing a new and unknown landscape. Britain, roughly south of a line joining the estuaries of the Severn and the Thames, was never covered with permanent ice. In the colder glacial periods, the vegetation in this southern region would have been of a tundra type; an icy waste followed by a burst of colour, consisting of dwarf willows, birch and pine, with some juniper; surrounded by mosses, lichens and rough grasses. In the warmer interglacials, there would have been the deciduous trees already listed, and flowering plants, such as buttercups.

Slowly, the vegetation cover increased over the whole of southern England. The mixture of tundra plants and increasing numbers of herbaceous plants, coniferous and deciduous trees, began to support roaming bands of Paleolithic people who were hunter-gatherers. Since the English Channel had not yet been formed, these bands could roam freely to and from present day France and the Low Countries, and beyond into the land which would become Europe. Between 14,000 and 10,000 BP there was uninterrupted tundra with developing coniferous and deciduous forests, right across into central and southern Europe. To Paleolithic man it would seem to stretch into eternity.

As the ice retreated northwards the tundra plants followed, colonizing the newly exposed rocky surfaces. These were joined by the herbaceous species of the established flora from the lands further south. As time passed, more plants from the milder southern refuges joined the gradual movement north, having sheltered safely during the 'storm' of the Ice Age. These plants may have been newcomers which had evolved during the millennia of the Ice Age, or they may have been simply natives, coming home to reclaim their territory. Before the ice came and covered the land, did the blue forget-me-nots, bright red campions and delicate pink mallows, bloom here, as they did in their recent Mediterranean home? Could plants such as the pimpernels, spurges and St John's Wort, also arriving from south-west Europe be part of the natives' return?

RELATIVE PROPORTIONS OF SELECTED POLLENS IN A PEAT CORE

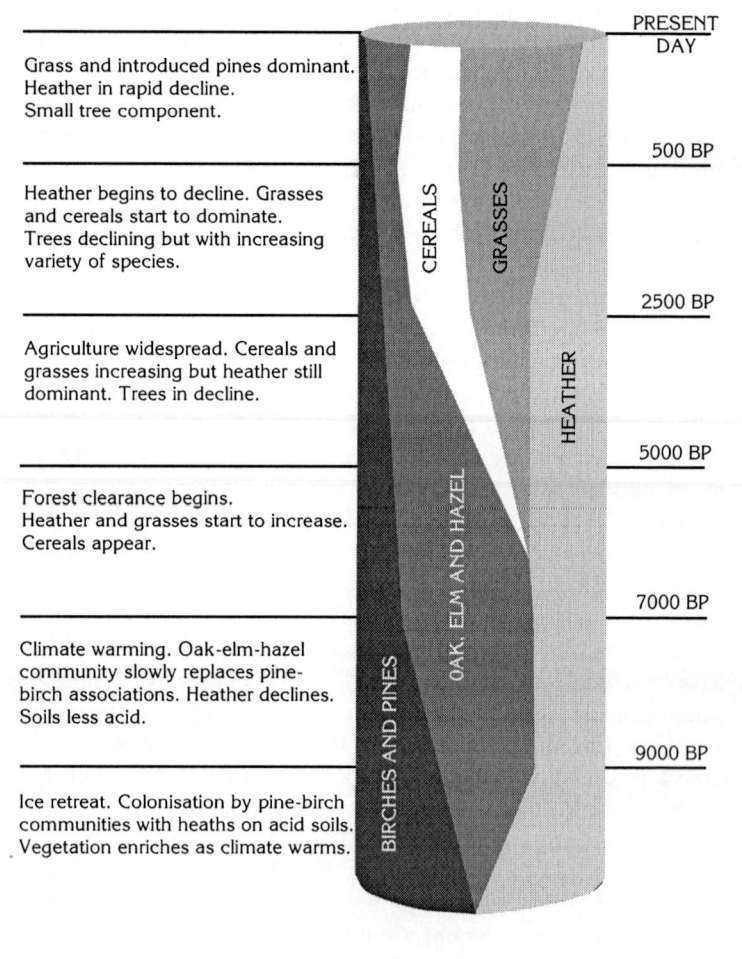

Grass and introduced pines dominant.
Heather in rapid decline.
Small tree component.

Heather begins to decline. Grasses and cereals start to dominate.
Trees declining but with increasing variety of species.

Agriculture widespread. Cereals and grasses increasing but heather still dominant. Trees in decline.

Forest clearance begins.
Heather and grasses start to increase.
Cereals appear.

Climate warming. Oak-elm-hazel community slowly replaces pine-birch associations. Heather declines. Soils less acid.

Ice retreat. Colonisation by pine-birch communities with heaths on acid soils. Vegetation enriches as climate warms.

PRESENT DAY
500 BP
2500 BP
5000 BP
7000 BP
9000 BP

Some plant historians claim that there were and still are native plants in England. Many plants considered to be very English such as bluebells, gorse, and the yellow Welsh poppy, spent the Ice Age in the mild Atlantic coastlands of present day France and Spain. They must have slowly moved north following the retreating ice. Skirting the retreating European glaciers, from their warm safe refuges in central and south-western Europe, came a different group of familiar plants, shrubs such as box, travellers joy, and honeysuckle, and herbaceous plants such as woad, lesser periwinkle, and various mints. It is debatable whether these plants were immigrants or returning natives.

The ice continued to retreat still further north around 12,000 BP, and all these plants and many more became part of the greatest 'rock garden' that the world has ever seen, extending from Ireland in the west to central Europe in the east. From the rubble and bare rock left by the retreating ice there arose a multi-coloured phoenix; an awesome display of colour produced by the advancing army of flowering plants, invading from the south.

In addition to the glacial till, the very fine material left by the retreating ice, more soil was formed by the action of the natural elements of sun, rain and wind on the rocks left behind by the glaciers. It was in these newly formed poor soils that slowly but surely the pioneer corps of scramblers such as blackberry, and climbers like old man's beard, cinquefoil, a frequent rock pioneer plant, along with knotgrass, a common ruderal, first became established. They steadily covered the land round the base of the rocks, growing in the newly formed soil. In the gaps around these pioneer plants and in the areas between the rocks, there was no competition for space, and new plant communities appeared. These were gloriously haphazard mixtures of trees, shrubs and flowering plants which would never be seen again in such natural profusion.

There were random clumps of the ubiquitous tundra trio of birch, willow, and Scots pine, along with juniper, in the loose soils at lower altitudes, where they were joined by ruderals such as chickweed, knotgrass and mugwort. The latter was to become a medieval cure-all for everything from moth protection to dysentery! The glacial till left by the retreating ice, had a stabilising cover of ribwort plantain—a familiar foe in many lawns—and docks. The docks today are characteristic of unimproved pasture, which is usually on fine soil like a lawn.

Here they stand out as isolated green or brown posts, since they are unattractive to horses and cattle who graze all round them. Yellow rock-roses and the bright blue flowers of both the cornflowers and Jacob's ladder, jostled with the pink thrift—sea pink—usually seen on the sea shore, a reminder of the bronze threepenny pieces and childhood summer holidays by the sea. The filmy yellow and white bedstraws covered the bare surfaces, as well as the pioneer corps of scramblers much as they do today, while the deep blue gentians and white mountain avens, with its feathery plume of styles, filled in the gaps between the rocks.

Some plants never made a complete return to England to join this celebration of colour. The strawberry tree, with its waxy white winter flowers and bitter tiny red berries, only reached Killarney in southern Ireland from its safe Atlantic refuge in present day France and Spain. The rising sea produced by the melting ice, cut off England from Ireland before it could travel further. The Lizard in Cornwall has plants such as Cornish heath which comes from the southern refuges and is found only in this region. East Anglia has an assemblage of dry central European plants, which got no further since the damper fens acted as a natural barrier to progress.

For nearly 2,000 years the land would be permanently covered in colour. Unfortunately, like all good things this riot of colour was not to last. Beween 10,000 and 9,000 BP the climate began to get warmer, and this heralded the arrival of the forest trees. Coming from central Europe, the Scots pine was the initial tree colonizer on the land which had become the 'Great Rock Garden'.

Eventually it formed the first forests on the land vacated by the ice. This conquest did not last for long, for as the climate continued to get milder, along came the deciduous trees such as oak and elm. They competed so successfully for space in the warmer climate that the pine woods retreated northwards, becoming a feature of the regions which later became the Pennines and Scotland. They took with them plants from the rock garden such as gentians, mountain avens and saxifrages, which preferred the cooler climate and still live in these parts today. The rest of the Great Rock Garden plants survived around the edges and in the clearings of the new deciduous forests.

This slow but inexorable take-over bid by the deciduous trees produced an overall cover of forest called the wildwood. The advance divisions of these trees from the refuges in the milder northern Europe included rowan (mountain ash), and wych elm, with hazel in the shrub layer and alder in the wetter lands. These were followed by lime and ash, with beech, yew and hornbeam arriving later on. The climate was getting progressively warmer and, as a result, the ice was melting causing a further rise in sea level. Just before the last land bridge connecting England to Europe disappeared, holly, and the wild service tree arrived around 7,500 BP.

The Tertiary period, beginning approximately 65 million years ago, was a time when the flora of the land now called northern Europe was tropical, similar to the rain forests of Malaysia today. After the last great Ice Age, which started approximately 750,000 BP during the Quaternary period, the flora became a temperate one, consisting of plants which arrived here from northern Europe, before the melting ice produced a rise in sea level and created an island state. Either of these two geological periods may be proposed as the time when a truly native flora evolved. Rightly or wrongly it is the post-glacial, temperate flora which has become accepted as the native flora although it was not the first to occur here. However it has become the starting point for the following periods in English plant history.

The name *wildwood* sounds evil and menacing, giving an image of the dark forests of our childhood nightmares, where forces unknown could hold us captive like Little Red Riding Hood. Yet this forest was a living thing, and as unthreatening as the sunny and captivating New Forest and Epping Forest are today. It covered the land like a large and colourful patchwork quilt, with numerous clearings like moth holes, through which the land poked from the uneven surfaces below. There were lime and yew woodlands in the warmer south. The north midlands had oak and elm trees with hazel in the shrub layer. South eastern England had areas of beech and hornbeam woods, while the Scots pine, the very early woodland colonizer, never left its refuges on the Pennines and in Scotland.

The water from the melting ice carved river valleys through this wildwood, exposing the earth, so creating places for flowering plants from the great Rock Garden, trefoils, pansies and the ruderals to become established. There was now room for damp area trees, the willows and alders to flourish. The regular flooding produced permanent clearings which slowly dried out as the waters subsided. Round the edges of these clearings, spindle and dogwood could bloom because they had sunlight. The increased light allowed the grassland plants daisies, vetches and selfheal to grow. As the temperate climate became established, so seasonal cycles appeared.

As a result, plants which flowered and set seed when there were no leaves on the trees, began to flourish, because there was more light. These included the bluebell with its sticky sap—glue for the feathers on future archers' arrows—daffodils and yellow celandines, the roots of which were considered to be a cure for piles. Later on, the red campion and cowslip, an early though unsuccessful remedy for Parkinson's disease, arrived and flourished. These would be followed in areas with increased light by the stinging nettles—those tough Ice Age survivors—willowherbs, foxgloves, with their cure for dropsy and heart complaints, and the cheerful yellow dandelions with their familiar clocks of seeds.

Animals also lived here and contributed to the opening up of the wildwood. Red deer migrated along regular routes which were possibly enlarged by aurochs and wild boar, both of which are now sadly extinct. Mesolithic man moved through the forest as well, creating his own trackways in addition to using those of the animals. He was a hunter-gatherer, so in the clearings he halted to pick nuts and berries as well as to develop his camp. He was slowly becoming a pastoralist.

Mesolithic man had discovered flints in the chalk and by using them in an axe-like tool he was able to cut down trees. He then enlarged these clearings to graze his animals and grow crops. Eventually these enlarged clearings would become the first grasslands and cultivated fields. As well as using the flints for tree felling and cleaning animal skins, Mesolithic man started to use them for fishing, in the numerous rivers, estuaries, pools and streams left behind by the melted ice. Around these fishing sites, the forests were slowly cleared to make the camps and contain the animals which were now part of his life. This was the moment for which many plants had been waiting.

Out from their enforced hiding in the dim light of the wildwood, came the plants which thrive on waste land. Wherever man settles he always produces rubbish. Here the rubbish was the remains of animal and fish carcasses, shrivelled fruits and nuts and pieces of unburned wood. Enter the goosefoots, so called because of their leaf shapes, stinging nettles—again—docks, because the land was so poor, black nightshade, and sheep's sorrel; all mounting successful claims for places on the rubbish dumps of Mesolithic man. Today in the twentieth century, they can be seen thriving on building sites and rubbish dumps of any kind, just as they did 7,000 years ago.

3
TILL, SOW, AND REAP

*Take not too much of the land, weare out not all the fatness,
but leave in it some heart.*
Pliny the Elder

The era of Till, Sow, and Reap covers the time from around 5,500 BP to the Roman invasion in AD 43. It was characterised by hunger-driven invasions, and the continual removal of the wildwood. The cleared land was used for arable farming and grazing, the first time that there had been organised agriculture. The technology of animal-drawn ploughing appeared, which slowly increased the rate of crop production and woodland clearance at the same time.

The first human arrivals were the European Neolithic farmers. These migrant people had steadily removed the trees from most of the southern European cultivatable land, and by repeated cropping had rendered it barren. With their herds of cattle, sheep and pigs they had slowly moved north from the Mediterranean area, looking for fresh pastures and land. Moving with them was an annual increase in wet weather and steadily decreasing food supplies. Eventually, in desperation, they set sail with their animals to a New World across the newly formed English Channel and North Sea. Here they hoped that they would be able to cultivate the land and herd their animals. The first Agricultural Revolution had begun. The clearings in the wildwood made by nature and Mesolithic man, were slowly enlarged by grazing and ploughing. The ploughs drawn by the cattle were antler horns, of which there was always a plentiful local supply. Later on, flints which were found in the chalk rock, were incorporated into the plough to increase the depth of the furrow.

Unfortunately the experiences in Europe were not remembered, or perhaps not understood? The Neolithic farmers did not manure the repeatedly-cropped land. Once it was exhausted they moved on and left it as bare soil. Areas such as the Breckland in Norfolk never regained any native plants or trees to cover the land. The damage was permanent, like many twentieth century farm practices. Even though the Neolithic farming practices had a disastrous effect in some areas of England, there were successes. For if a farmer's job is to produce sufficient food for an expanding population, he was excellent at his work. The large armies of men who built the henges such as Scarra Brae on the Orkneys and Stonehenge in the south of England, would have to be fed. Modern estimates suggest that two million man hours were needed to build Stonehenge with its twenty-six tonne sarsen stones. The food supply for a workforce averaging 54 inches in height must have been enormous and continuous, and is hard to visualise.

A less successful farming practice was the use of elm for animal fodder. When there was no pasture, the animals were fed on elm leaves from the abundant trees in the wildwood. There is evidence of pollarding, and the use of these pollards in the trackways across marshy land, notably in the Somerset Levels. Perhaps this use, combined with the effects of elm disease, produced the decline of the elm population from which it has never recovered.

On the positive side, the Neolithic farmers brought about a grassland flowering nearly as bright as the Great Rock Garden. As the farmers cut back the wildwood for space to graze their animals, out came the grasses and light loving plants, held for so long in the shady woodland. Grassland plants such as buttercup, daisy, dandelion and trefoil appeared in the grass, to be munched by the grazing mammals. The longer grass round the edges would have been communities of bed-straws, vetch, lesser knapweed, and field scabious. These were the early grassland communities, whose descendants can be seen today.

But this was not all that flowered. Neolithic man brought with him the seeds of several different plants, common in our flora today. Cleaning or screening seeds before sowing is a present-day practice. In those far off days, just having sufficient seed left over for next year's sowing would have been an achievement. The farmers sowed seed initially from European sources.

Enter the primrose. This delicate flower so typical of the English spring, bloomed and flourished in the mild, damp climate, spreading out over mainland Britain. It had declined since twentieth century man started to dig it up, transplanting it from woodlands and hedgerows, usually unsuccessfully, into suburban gardens. Today as the millennium approaches, it is increasing its distribution again; this time along the verges and banks of the motorway network. Here it is safe from the collector's trowel in its new twentieth and twenty-first century plant refuges. Curiously it seems immune to the diesel and petrol fumes which pollute the air along these highways. It is also re-appearing along many a country lane, where a decrease in local authority spraying and a heightened awareness of the need for conservation are having their effects.

Two plants which arrived by courtesy of the Neolithic cattle—in their bowels—are charlock and common fumitory. Charlock was and still is a pest of cornfields, though to the casual observer it gives a lovely splash of yellow at the edge of a field. Today it is a serious pest to farmers, and is termed by Grigson 'a vegetable rat'. It will come up year after year, though selective weedkillers have now sounded its death knell. Its sole offence is that its seeds contaminate the corn crop at harvest time. Common fumitory is an inconspicuous plant by comparison with charlock. It has delicate pink to mauve flowers and is often found round the edges of fields and on pathways. The name refers to the fact that the juice of the plant can make the eyes run, as they do when we are near smoke. Another arrival, probably with the grain, was the white campion. It is as harmless as fumitory, and is found in a variety of places, especially on verges near cultivated land.

The successful farming techniques of Neolithic man had the inevitable effect of attracting more invaders. The news of good agricultural land travelled fast, then as well as now. The climate here was becoming warmer and drier, increasing the crop yields. Enter the Beaker People from present day Spain. They brought skills in metal-working to the Neolithic culture, as well as their grains—wheat and barley—which were not cropping very well on the increasingly hot Mediterranean lands. These grew profusely in the south of England and must have added food as well as manpower for the building of Stonehenge. Again this was not all that arrived, for in the grain came poppy seeds. For the first time the bright red poppies would bloom in the cornfields of southern England.

They have always been associated with corn and today we can record a poppy-corn culture starting with the Assyrians who called the poppy 'the daughter of the field'. The Romans had Ceres their goddess of the harvest, carrying poppies and corn. In the twentieth century Monet painted fields of poppies and corn. It is still a symbol of life as we, the living, remember the dead of many wars with our paper poppies. Today, in spite of extensive grain screening, the digging of trenches for gas pipe lines, telephone cables or drainage pipes can give a vivid red slash of poppies across fields and along verges. These flowers are reminders of early European invaders, and their seeds' longevity, which is at least fifty years.

During the warmer drier climatic phase, along with the Beaker People, Bronze Age man arrived from Europe. Like the Beaker People, he was probably looking for land on which to settle, grow his crops and raise his animals. The rate of wildwood clearance increased, as these invaders settled down to life in England. The warmer, drier climate encouraged the removal of the forests on the higher land, like Dartmoor, so that it could be used for grazing. The floors of the valleys round the higher land were also cleared of trees, becoming arable lands or heaths, much as they are today. Probably half the wildwood had been removed by this stage, for agricultural needs, roads and settlements.

The population had been steadily increasing, and would continue to do so for some time to come. Apart from the removal of the wildwood, Bronze Age man made his mark with the plants which he brought. Black bindweed, bristly ox-tongue, fat hen, Good King Henry, pennycress, and the dull red deadnettle, all arrived at this time by several different means. Black bindweed grew with the grain—it probably always had done so—the starchy seeds adding bulk to the grain harvest. Today it is a rampant though inconspicuously flowered plant on waste land, around cultivated fields, on open grounds and tracks. Ox-tongue, with its basal rosette of prickly leaves is another plant of waste land, probably arriving in the grain like black bindweed. Fat hen and pennycress probably reached here in the bowels of Bronze Age man's cattle. They were probably distributed round the land via the rubbish dumps and animal feeding grounds.

Pennycress is the scourge of many gardeners, producing large numbers of seed which are readily dispersed. Fat hen, along with Good King Henry, relations of the goosefoots, have highly nutritious seeds rich in albumen and fat. The plants were a regular supplement to man's diet at that time, unlike pennycress which apparently had no use at all. Along with these plants, by method unknown, came the red deadnettle. Today it is a small, shade-loving plant common in hedgerows and on the edges of woodlands. Many hybrids of red deadnettle can now be seen in garden centres. The basic genetic stock is very stable and the results are numerous attractive varieties of bedding plants, from an inconspicuous Bronze Age arrival.

Usually the source of information on plant arrivals and usage is pollen analysis. Occasionally there are other more unusual ones. Tollund Man, also known as Pete Marsh, was thrown into a Danish peat bog—between 400 BC and AD 400—apparently strangled by a rope found round his neck. His body was completely preserved, even the contents of his stomach were identified from seeds. The condemned man ate a hearty breakfast of black bindweed and fat hen gruel, along with barley, linseed, corn spurrey, and heartsease. Unfortunately, there are few such interesting and unusual sources of plant information. But the food does not tell us what sins he had committed to merit the death penalty; so we will never know.

Any story of these islands will always include changes in the weather. Over a small area, even a slight alteration in the climate will have an effect. When around 2,500 BP the climate became cooler and wetter the results were disastrous. The thin soil, on the tree-cleared highlands, soon lost its mineral content through leaching and became increasingly infertile. On this land, now deficient in minerals, peat bogs appeared, first on the upland pastures and then on the arable fields.

From late Bronze Age onwards, places such as Bodmin Moor, Dartmoor, and the Scottish islands of the Orkneys and Lewis were slowly covered in peat bog. The re-colonization of these bogs by juniper—again—bog myrtle and the ubiquitous bog cotton was no compensation for the lack of arable or grazing land. Communities could not grow enough food, and people living there starved to death.

Towards the end of the drier period, around 2,800 BP and at the start of the wetter climatic phase, there was a new phase of invasion. Celts from the lands which are now the Low Countries slowly invaded England. They probably came for the rich farming land, developed by the Neolithic and Bronze Age farmers. These people had iron-making skills and with their iron axes, trees were felled more easily than with the earlier flint-based tools. Smelting the ore for more axes and ploughs needed charcoal—more trees gone—so the wildwood was reduced even further. With their iron ploughs, the energetic Celts were able to cultivate the last, so far untouched, lands of heavy clay soils in the southern valleys and the plains of central England.

In southern England the Iron Age teams of oxen ploughed up the Neolithic grasslands and removed the woodlands, until some areas of southern England were just fields of waving corn. Slopes, once covered by beech woods, were lost to crops such as linseed—flax—and a form of pea, as well as barley. The north and south Downs may never have had a full tree cover, even when the ice first retreated. Here Iron Age man grazed his sheep creating the first downland turf as we know it, rich in native flowers such as bird's foot trefoil, rock-rose and native thyme,

The political stability of southern and central England was maintained by hill forts. These were places of safety into which these Celtic invaders retreated when the natives became restless. It was not surprising that there was unrest, since the local farmers had lost their land to the Iron Age ploughshares. Today these forts look out over a landscape which bears no relationship to the one which was present when they were constructed. At this distance in time, one wonders how these forts, such as British Camp and Maiden Castle, were constructed without earth moving equipment.

All this political stability had an unexpected downside. The increased population had to be fed, so more land had to be brought into cultivation. The change in the weather made this more difficult, since the land was rapidly becoming leached and so less fertile. In addition to the gentle slopes, steep land was also cleared of trees and by cutting into the hillside, terraces were created. These new narrow strips of terraced land are called 'lynchets', and can still be seen today. They look like giant contour lines drawn round the hillsides in areas of Dorset, Hampshire and Sussex. They are now sadly devoid of trees and few other plants grow there, since the soil was irreparably leached in the increasingly wet weather. They could never recover the original plant communities.

During all this increase in cultivation from Neolithic time onwards, what was happening to the grasses and flowering plants of the Great Rock Garden? On reflection, it is amazing to realise that for many plants, it was a return to the land, just as it was when the ice was retreating for the last time. There was a bonus too; it was warmer. The plants, like an invading army, reappeared from the wildwood where they had been contained since the trees had first taken over the land. They slowly took over the stony yards and rubbish heaps round the settlements, and became established in the soil on the land exposed by cultivation.

The early heaths made by felling trees or the extension of the natural gaps in the wildwood, became refuges for acid loving plants just as they are today, often quite untouched. It is here that tormentil, heaths, heathers, and lousewort thrive, to give us those vivid flashes of yellow, purple and mauve colours. The grass meadows so lovingly created; the cleared land around rivers for animals and paths were colonized by the plants which had lived under those conditions at the end of the Ice Age. The downlands already mentioned had communities of plants thriving on the regular grazing. Due to the steadily cooling climate with increased rainfall, the woodland was regenerating on southern lowland sites which had been abandoned.

However the picture in the north of England was very different. The land was less hospitable with high mountains and steep-sided valleys, the climate was colder and the natives very unfriendly. In fact the subjugation of the south had taken the Celts some time, so the northern lands had been left in peace. Like the Romans who followed them, the Celts did not feel that the land had immediate appeal on any count. A mountainous land and a wet cold climate created very different meadows from the southern grasslands, created by Neolithic man. When Iron Age man finally felled the trees, the flora which appeared to cover the land was very different from its southern counterpart. Some of these plants were interglacial species. Amazingly, examples can be seen today in the rare assemblages of plants in Upper Teesdale. There are gentians, sandworts, shrubby cinquefoils and Teesdale violets, flowering as brightly as they did several thousand years ago.

In this region, Iron Age man had to develop a pattern of farming suited to this northern climate, which had to include a hay meadow. This provided the cattle with fodder in the long winter months, when there was no outside feeding. The meadows were never grazed, and the grass was cut, and still is mown once the plants have set seed, usually in August. This controlled cycle of cutting and harvesting ensured a good crop of grass in the following year, as well as winter feed. Yellow rattle whose seeds make a noise in the dry pod, blue meadow cranesbill, the round-headed yellow globe flower, the white flowered pignut and the bright red betony, give these hay meadows a brilliant haze; a strong contrast against the grey and white stone walls of the northern region.

These meadows, around the edges of the Pennines and in the Dales, have changed but little over the centuries. Today, the flowers bloom as they did when Iron Age man, having removed all the trees, finally let the flowers of the meadow back into the sun. They flower in splendour, as well as feeding the livestock. Here, as in southern and central England, the plants of the Great Rock Garden had returned safely to cover the land with colour once more.

The last half of the Iron Age was marked by a slowly cooling damper climate. This had initially aided man's agricultural activities, for plants grow well in damp moist conditions. Unfortunately, it was not enough to feed the rapidly increasing population, even with the increased clearance of the wildwood and the creation of lynchets. The climate continued to get cooler and wetter, making the land increasingly infertile through leaching.

So, by the end of the Iron Age, which overlaps the Roman era, there had been a steady decrease in fertile farming land and an increase in the moorland, which reduced the available farmland even further. Obviously this reduced the already lowered crop yield, resulting in starvation and high levels of mortality in places such as the Orkney and Shetland Isles. Man had created the moorland by cutting down trees for land, on which he could graze his cattle, so feeding the increasing population. This huge mistake reaped, and is still reaping, a very bitter harvest. Due to the nature of the soil which was formed, it has never been possible to convert it back into arable or useful grazing land. This point in time can be described as the First Biological Crisis, as shown in the diagram of the Ages of Plant History on page five.

To end the period on a positive note, Iron Age man added three notable plants to the English flora; fool's parsley, black mustard, and borage. Fool's parsley, whose resemblance to the real herb is excellent, arrived with a cousin of the dreaded charlock, black mustard. Both these plants were probably used as condiments to spice up the fat hen and black bindweed gruel. The black mustard had, and still has today, the properties of the family, that is to say a fiery taste and being a good ruderal, it would thrive in most weathers and in most places. It was a most convenient herb to grow in the garden.

Borage with its star-shaped blue flowers and grey-green shaggy leaves, appeared towards the end of this period. The Celtic word 'borrach' means gladness and courage and the Latin 'burra' means a shaggy garment. This suggests that it might have been a Roman introduction as well. Infusions were made from the whole plant and were taken to relieve depression and as a fever cure. Today's uses are more mundane in salads, soups, and as a garnish for the popular summer drink Pimms. Does this drink relieve depression and reduce fever?

So the Iron Age with its decreasing population and advancing moorlands drew to a close, leaving the Romans to take over this land called England. Up to now the flora had been slowly building up, conditioned by the climate and the activities of early man. The arrival of the Romans was to produce a profound change. During their 400 years here, they created a level of internal organisation which was not reached again until the Industrial Revolution of the eighteenth century. The damage to the land was permanent and could never be repaired.

Borage

4
VIATORES

Veni, vidi, vici.
Julius Caesar

By 2,000 BP England had become a rich nation, exporting grain and metal goods to mainland Europe. At this time the Roman army was in Europe, struggling to maintain order in Gaul and so permanently establish Julius Caesar's political credit in Rome. Napoleon said that an army marches on its stomach, and this was just as true for the Roman armies many centuries before. Once the occupying army had used up all the local supplies, they began to look elsewhere for a regular supply of grain. Once again hunger was to drive invaders from Europe to England.

The childhood rhyme 'Romans came across the channel, All wrapped up in tin and flannel', came true in AD43, after some preliminary skirmishes to check out the best landing places. However this was no pantomine act as the rhyme might suggest, but a full scale invasion. For three hundred and fifty years afterwards, the Romans developed a robber economy here in England not unlike the early phases of European colonial rule, during the eighteenth and nineteenth centuries on the African and Asian continents.

One reason why the Romans were able to conquer England so quickly was the well-developed road system. Mesolithic man had established routes in the wildwood connecting places as far apart as Norfolk, the Lake District and Cornwall. Neolithic man who followed had enlarged this communication system with their trade in flints and the Bronze Age metal traders had increased the roads still further. The Iron Age culture had added even more routes as they developed a trade in hides and grain, both at home and in Europe.

They returned from the continent with luxury goods such as wine and pottery, to be distributed on the ever-expanding road system. So the Roman Highways Department was not solely responsible, for those straight red lines in the road atlas. They did add some of their own roads later, as a response to local features such as river crossings, and for strictly military purposes. This was probably to the great disgust and fury of the local tribes, when the Romans went across their crop and grazing lands. Unfortunately the locals would have been no match for the swords and spears of the Roman legions, and there was no Court of Appeal or Land Tribunal in those far off days.

Overall the Romans inherited a well integrated system of supply routes for the initial invasion force of 40,000 men and more legions later on. Between AD43 and AD81, when most of England and Wales was being brought under Roman rule, the road engineers improved the existing roads and extended them to suit their own purposes. There were roads connecting Richborough, the original invasion port, to Caenarvon in Wales, Exeter in the south west and the Antonine wall on the Scottish border.

The roads were raised on an agger bank, with stones dug from a ditch at the side. The troops marching along these roads, needed feeding and so did their horses and mules in the baggage trains. The sudden influx of men and animals was always going to be too much for the local food supplies to match. In a good year the volume of grain in surplus would not have fed the invading army. As a result they brought with them from Europe food for the troops and fodder for the animals. Inevitably some of this food and fodder, left behind at the numerous staging posts, would contain seeds from their place of origin in Europe.

Enter the next wave of conquerors; a group of plants which, quietly and with great efficiency, successfully invaded the agger slopes and ditch banks. It has been estimated that these slopes provided up to six thousand acres of bare surface ready for immediate colonization by these European seeds. In an alien environment, they were probably less competitive than the native plants, but they were there before the locals arrived. These new plants would be suited to exposed stony ground, since many had a south European origin. In addition the damper and less extreme climate would increase the inital growth rate, making it harder for the locals to compete successfully.

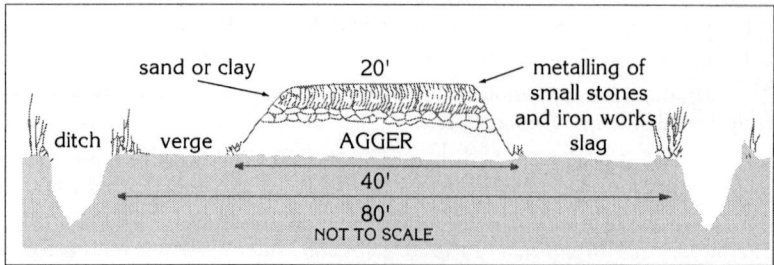

Surprisingly, there were other modes of transport for these invading plants from Europe. Over many years, the muddy sandals of the Roman soldiers and the hooves and hair of their animals would have helped move more of these seeds along the roads. The footwear of the local traders, who followed the soldiers, selling skins and pottery, would also have contributed; we were a nation of shop keepers long before Adam Smith and Napoleon gave us that name, and unknowingly these market traders were doing more than just selling their wares.

The newcomers on the agger and ditch banks were either plants of waste land, cultivated fields or those associated with grain. Other European arrivals such as swinecress still live on stony paths, along with the scarlet pimpernel, with petals which open around eight in the morning and close around three in the afternoon, unless it rains, when there is early closing to suit the weather. Another arrival was sowthistle with prickly or smooth leaves, yellow flower heads and silky parachutes of seeds. It is common today on waste ground and roadsides, just like the agger and ditch surfaces of the Roman period. The unpleasant smelling field woundwort with its dull red flowers is another present day inhabitant of waste ground, which arrived here from its home on the dry places in southern Europe.

We know from soil analysis around Roman sites, that the flowers of the cornfield and grasslands which arrived with the Romans were numerous. Some of these have become rare with time, due to modern grain-cleaning techniques. One of the well documented plants is corn-cockle. This has been screened out because of the bitter taste given by the seeds, and their poisonous effect when used in bread in large quantities. Corn marigold with its bright yellow daisy-like flowers, was once very common. Its name is still seen in places like Goldhanger in Essex and Golding in Shropshire. This plant of Asiatic origin has been eliminated because it had take-over tendencies. The rough-headed poppy is still found at the edges of cultivated land. Along with field gromwell, they are both apparently immune to twentieth century sprays. Corn spurrey, another arrival, was used subsequently as a food crop for man and beast, grown on poor soils especially in the north of England.

Perhaps the most interesting plants which the Romans brought were the deliberate introductions which have escaped over the garden wall, becoming vagabonds and naturalizing within our flora. Unlike previous invaders the Romans were a literate and numerate people. They recorded what they saw and what they did, which has been to our great advantage. The earliest plants which they brought were probably grown in the infirmary garden. Soldiers, as well as administrative staff, would fall ill as a result of injuries received in battle, or due to the cold unforgiving climate in which these Mediterranean people found themselves.

Since it would be too far to send home to Rome for a repeat prescription, many garrisons developed infirmary gardens. Seeds brought from the camps in Gaul would have been grown to produce painkillers, cough cures, sedatives and anti-depressants. Henbane, opium poppy and mandragora were grown to produce painkillers, although the Romans admitted that they could not adequately estimate the dosage. One surgeon suggested that indifference to the patient's cries was better medicine than a draught from any of that Great Trio of painkillers, which could be fatal. It is an intriguing thought that if the Romans had not brought and recorded the effects of henbane, Dr Crippen may not have used it to dispose of his wife and other female victims around 1910, denying them old age and a natural death.

Opium has been used as a painkiller for over five thousand years, and has threaded its way through British history from the Roman period onwards. The Sumerians had learned how to extract the opium juice from the poppy capsule and handed on the skill to the Egyptians, the Arabs and the Indian sub-continent. The Greek physician Hippocrates, the father of modern medicine, recommended opium as a pain-killer as did the Roman physician Galen, who used an opium preparation called mithridate. In northern Europe, the opium poppy capsules grow small and often yield very little juice.

From the Roman period onwards, the whole plant including capsules and seeds was boiled to give an infusion, which yielded a low level of opium. Although the medieval Catholic Church disapproved, its use continued. By the eighteenth century, most homes had some opium-based product, always seen as a medicine. We were a rich trading nation by then, so the source was not always a European infusion. Opium imports alone from the Far East produced huge profits for the British East India Company. Unfortunately it has always been addictive and with increased availability quickly became a recreational drug.

It is probable that Coleridge's fantasy dream poem, *Kubla Khan*, was written under the influence of opium and Dr Arthur Conan Doyle clearly describes the use of opium by the detective Sherlock Holmes. The laudanum of the nineteenth century, an opium-based medicine, has a well-documented history of addiction, particularly by women. It can be argued that the Romans started opium addiction in northern Europe; but that is a matter of opinion.

Collecting opium

Another Roman introduction into the infirmary garden was hemlock. It is definitely a pain-killer, but it has so many unpleasant side effects one wonders why it was used. Among the general symptoms of hemlock poisoning are excessive salivation, diminished breathing frequency, irregular heart action, convulsions and finally total paralysis. Around 400 BC the Greeks used an infusion of this plant to put Socrates to death. It must have been a dreadful and painful end. Knowledge of these effects was not lost when the Romans left in the fifth century AD. John Keats, poet and doctor, refers to them in his beautiful poem *Ode to Autumn*:

A drowsy numbness invades my senses as though I have of hemlock drunk

As a doctor he would have understood the early effects of hemlock extract. Henbane, opium poppy, and hemlock, vagabonds all, grow freely in our flora, making them perhaps the first and the best-documented of the Roman denizens. Mandragora did not survive in the wild, it was probably too cold.

Other Roman plant introductions, whose extracts stocked the garrison pharmacy and now grow wild in our flora, are the greater celandine, ground elder, balm, and white mustard. Greater celandine was a mild sedative and easier to use than the Great Trio, as well as being the basis for an antiseptic eye wash. Ground elder produced a supposed cure for gout, an unfortunate side effect of the numerous lavish banquets. Today, as well as living in the wild, it has returned from whence it came—the garden. Here it is not a cure for gout but a prescription for back ache, since it is very hard to dig up!

Extracts of balm were made up into a tonic which was used in a liqueur-like drink, to relieve depression and anxiety. These are not surprising ailments, for the permanently damp and grey weather must have come as an enormous shock to people who were used to hot summers, with long hours of sunshine. Balm now only grows in the milder climate of south east England, which is more like its native land. White mustard produced large amounts of small seeds which were ground up and used as a mild laxative, not as a condiment. This plant is a true ruderal and grows vigorously today with its cousins charlock and black mustard, on any waste ground which can be found.

Another denizen was the sturdy, yellow-flowered alexanders which moved out of the herb garden, not from the infirmary plot. In the days before refrigeration, because fish would go bad very quickly, the Romans brought alexanders to cook with it and hide the taste of the tainted flesh. It has settled near ponds and old castle moats, where it is still associated with fish. Pennyroyal, a less common Roman introduction and easily confused with other native mints, was another herb brought to England to disguise the flavour of tainted food, usually meat. The strong minty taste also helped to cover up the effects of the lead cooking pots, used in Roman kitchens.

These utensils produced a metallic taste in the mouth which would disguise the flavour of the food, unless it was very strongly flavoured. Using this herb and probably native mints as well, the Romans claim to have invented mint sauce; well, they certainly had the vinegar or sour wine to help the sauce along. Today pennyroyal is less common, due to the drainage of farm lands, having a limited distribution around farm ponds. The strong black mustard, which arrived here with Iron Age man, was often used with milder Roman white mustard, of southern European origin, to create the same flavouring effects.

Moving out of the gardens, and into the wild with the vagabond army of medicinal and culinary plants, were the sweet chestnut and walnut trees. Sweet chestnut trees had been brought from the Mediterranean for two reasons. Firstly, the nuts stored well, providing food in the winter months for the animals. Secondly, the wood split easily and was extremely rot-resistant, therefore excellent for fence posts and out-buildings. Stands or single trees are still found near the sites of Roman settlements, descendants of the Roman introductions. Walnut was used to provide wood for domestic purposes, since it had a dark and decorative grain, much admired today in quality furniture. Like sweet chestnut, the nuts stored well and were a useful addition to the winter diet. It is more susceptible to frosts, so it is only found in warm sheltered places, often on limestone, and is usually solitary.

Did the Romans bring the sycamore tree to provide fuel for their numerous hypocausts? This rapidly growing tree is from central and southern Europe and not the Mediterranean, from where the main Roman introductions originate. Unfortunately soil or pollen evidence is sparse and inconclusive. The first valid written evidence is found in Lyte's *Herbal* of 1587. The sycamore may have arrived by chance more than once; no one seems to know and the plant historians cannot agree. The case rests.

Whatever happened to the large number of Roman introductions which did not become permanent denizens? Herbs like sweet basil, bay, caraway, coriander, fennel, sage, and thyme were brought for culinary use as well as their medicinal properties. Lovage, hyssop, and pot marigold were Roman introductions, as well as the elegant bear's breeches. The answer is that these plants remained safe, growing in successive generations of monks' gardens, physic and botanic gardens, cottage gardens of the poor, and the formal gardens of the rich.

Although caraway and bear's breeches, along with fennel, have made a limited bid for freedom.

They were, and still are, poor competitors for space in the wild, and their numerous uses made them in many ways, a type of protected species. The aromatic plants like lavender, whose oil cured headaches and insomnia; rosemary with oil for rheumatism and heart complaints; the strong smelling rue used as an insect repellent and general stimulant, are still with us in these safe garden refuges. In many ways they have become part of our history. The Lord Mayor of London carries an early example of preventative medicine at his procession; a small posy of rue to ward off the plague, as well as the smell of the common people. Elizabethan lutes were made from lavender wood because the sound that was produced was the sweetest one. Without lavender, would we have had tunes such as Greensleeves?

Then, as now, man liked the security of insurance. For the Romans houseleeks were an insurance against fire, especially from lightning which they attributed to the gods Zeus-Jupiter *et. al.* These tiny plants travelled across Europe from the Mediterranean to England to be planted on the roofs of Roman houses and barns. Their bid for freedom took them to old stone walls around fields and gardens where they thrive today. In Somerset they have a more unusual name, *welcome-home-husband-though-ever-so-late*. Houseleeks became an insurance against fire—and late husbands? The Pilgrim Fathers took them to the New World, where they escaped into the wild for the second time. They grow on the stone walls of the New England States today, elegant reminders of a Roman insurance policy.

The last Roman introduction, perhaps the most familiar, is the carnation. From the Atlas mountains of Africa, the carnation travelled through Spain to Rome where it was called the Spanish flower. The present-day name refers to the Roman use as a ceremonial flower; *corona* means coronation or crown. It has sometimes been called clove flower or gilly flower due to its smell and probable value as a spice. From Chaucer's *gylofre* and Shakespeare's gilly flower, the carnation is found in every florist shop and motorway service area. The bright frilly flowers of innumerable hybrids have become our nation's most popular floral gift. Yet it never made it over a Roman garden wall and into the land beyond, to join other vagabond-denizens enjoying the new freedom. We will never know why.

The final withdrawal of the Romans at the beginning of the fifth century was both a historical and a cultural landmark. It was also a relief, for it gave our flora a much needed breathing space. The Romans had irrevocably changed the landscape by clearing woodland, draining swamps and marshes, altering river courses, intensive farming practices and the building of cities and towns with their connecting roads. In addition they had left their mark on the flora through their own plant introductions.

Time was needed to allow the new plant communities to evolve together into a stable flora within the new ecosystems which had been created. This stabilisation, both natural and cultural, took place during the calm of the Dark Ages and through the Anglo-Saxon period, ending at the Norman Conquest. From then on changes were more gradual and less spectacular. The intensity of environmental change in England was not to be seen again until the second half of the twentieth century, called by an eminent plant historian, Oliver Rackham, *The Locust Years*. Not even the Industrial Revolution with all its satanic mills, created such permanent changes and lasting damage to the English landscape.

Scarlet Pimpernel

5
THE DARK BEFORE THE LIGHT

The bright day is done and we are for the dark.
William Shakespeare

The period from AD400 onwards is called the Dark Ages for a good reason; we know little about it. The information which we possess was gained by looking backwards from the Anglo-Saxon period, which ends conveniently at AD1066. Why did the Roman civilisation start to decline? At the end of the fourth century, the Romans were attacked by the Scots from the north, the Irish from the west, and the Saxons from north west Europe. The Roman Empire was already in decline, so could not supply this northernmost province with sufficient troops to hold back the invading tribes.

Eventually, due to pressure from Rome, by AD407 even the small British garrisons were withdrawn totally. By AD410 the withdrawal was formalised and we were on our own. The society, communications, trade, even government collapsed, as local political units spent more time fighting one another instead of the invading forces. The whole fabric of *Pax Romana* did not break down completely, with murder and sudden deaths wiping out the population. If it had collapsed, within fifty years much of the cultivated land would have returned to scrub and woodland, with hedgerows and walls becoming lost inside them. This did not happen.

So what were the people doing during this time called the Dark Ages, and who were they? Many would have been part Roman, descendants of the civil and army personnel. Apart from the invading tribes, the rest of the population would have descended from the Iron Age people, here at the time of the Roman invasion. Most would have been influenced by the *Pax Romana* in some way and have had a place in the society of the time. These people probably developed the forerunner to the open field system. This would have been a practical solution for small landowners, for by combining their lands there would be no fences or hedgerows to maintain.

There is evidence that they moved into villages, instead of living in the isolated hamlets and farms of the Roman period. The Romans had planted hedgerows, which appear to have been extended, and more were planted as land use changed. Since there were fewer people to feed, some of the arable land would have returned to grassland. So, not for the first time, out of the woodland came the plants to reclaim some of their old territory, as well as find homes on the new hedgerows and pathways of the villages.

The first Saxons, arriving from the start of the fifth century into the early Dark Ages, were an aggressive and land-grabbing people who left no written records. The second wave of Saxons, arriving in the sixth and following centuries along with the Angles, were literate farmers. They came to find new lands to farm; not the first time that England has been invaded for its food potential. Their written records of land deeds and claims do not show a land completely wooded, or devastated. It was not a new landscape, but an established one with roads, lanes and fields. They appear to have farmed more intensively than the Romans, both in the woodlands and on the open land, producing more crops to feed an increasing population. As the small political sub-units grew larger and finally became Kingdoms, internal trade began to reappear, such as the salt routes from Droitwich and Nantwich. This started the development of a new road network, unlike the Roman one, since the needs were different.

The Romans conquered a land growing bread wheats, barley and oats. During their occupation, rye was added to the cereal list, which was to become the main peasant cereal in the Middle Ages. It was the easiest cereal to grow and overall the best cropper. However the Anglo-Saxons grew mainly barley and oats with a little rye. It is the plants which are associated with the cereal crops which thrived so well. As a result of the intensive agriculture they spread easily, since their seeds are not readily separated from the grain. A good example is corncockle, which first appeared with the rye the Romans had introduced. It spread rapidly, since rye is a less demanding crop than the other cereals, so was grown over a wider area. It became a plant associated with bread wheat and so spread even further.

Another Anglo-Saxon crop which had plant followers was flax. It is not known whether the Anglo-Saxons grew this crop for food, or for the fibres to turn into cloth. It is not even clear how flax arrived here in the Neolithic period, nor whether it was deliberately grown for food or clothing. It was growing in Europe at the time of Tollund Man, since the seeds of linseed were found in his stomach, so in this case it was obviously a food. Linseed, the fruit of the flax, is rich in oil and protein and so would be a rich food source. Since early Biblical times, flax has been a source of fibres for producing fine cloth. It needs a very fertile soil, especially if grown for the fibres, not just the seed. For reasons unknown, pennycress, a Bronze Age arrival, and corn spurrey, a Roman period arrival, are plants always associated with flax. They spread rapidly across the country with the crop, at the same time as corncockle, for the same reasons. Today, both field pennycress and corn spurrey are common but the delicate blue flax is a rare relic of cultivation.

Hemp, another Roman introduction, was grown by the Anglo-Saxons. A native of the Far East and China, it was grown to make ropes and coarse cloth, by the same retting, or soaking, process as flax. The craft was taken by the Pilgrim Fathers to the New World, for the same uses. The Boston ropeways were made of hemp fibres, as well as much of the early clothing. Hemp became a vital crop in the American War of Independence, since there was no trade with England. As time went on, hemp or cannabis, being a natural vagabond and liking its new home, moved out into the lands beyond, where it spread rapidly as a denizen over most of the eastern seaboard. It became a rich source of the hallucinogenic drug, marijuana. In England it never strayed out of the garden but it has become a visitor, whose presence is illegal!

When the Romans left, large areas of the four million acres under cultivation began to revert to shrub land. This cut down on the light which was needed by many plants at soil level. These were the plants which had left the wildwood when the land was first cleared. Now they were on the move again, perhaps for the last time. Plants such as goosegrass, fat hen, pennycress, bindweed and plantains were relegated to the edges of paths and clearings, since they needed light. The communal field system developed by the Saxons was a bonus for these plants, which have lived around paths and roadsides ever since, in a well lit world of their own. Broad-leaved plantain, found today beside all unmade paths as well as road verges, was known as waybread by the Anglo-Saxons, due to its regular presence on the paths and ways. It was resistant to trampling then and now.

Through the Dark Ages and Anglo-Saxon period, and into the Middle Ages, the dominant woodland tree was the pedunculate oak. Full grown trees were rarely felled, since they were too big to move or cut up. The wood needed for building was produced by coppicing, which ensured a continuous supply. This had another benefit, for it gave light to many woodland shrubs and plants, which have survived to the present day. An excellent, though accidental, conservation method developed, due to an inability to fell big trees and carry them away.

Quietly growing and surviving well in the monastery and abbey gardens, were fruit trees, cultivated species of apples, pears, cherries, plums, medlars and quinces, many of them having been brought here by the Romans. Their influence was also seen in the raised beds, in which grew the edible or medicinal plants; alexanders, leeks, opium poppy and greater celandine. The vines growing on the walls were reminders of the Roman occupation. Apart from the wine, there was also its use as an antiseptic and as something to add to water to make it free from infection. Clean water was rarely available in those days, for even well water was unreliable. Two very decorative garden survivors of the Roman period are the Christmas rose and the beautiful Madonna lily. The Venerable Bede, an eighth century theologian, viewed the latter with great pleasure. Did he know the odd use which the Romans had for this originally Greek plant? It was a source of ointment for the corns raised by the rubbing of sandal straps. An odd use for a lovely plant, but its success rate is not recorded.

The Christian communities became centres for the exchange of information about the cultivation and uses of herbs. These were the only places of learning with people who could read and write. Plants like birthwort, with its abortive properties, and peony with its anti-asthmatic and anti-epileptic properties arrived here in Anglo-Saxon times. Both plants are natives of south east Europe and like many European plants before it, the peony has remained in the safety of the garden. These two plants are rarely used now by western herbalists, for they are considered to be too extreme in their effects. Today, the peony blooms in our gardens giving great pleasure as a reminder of the early monks' infirmary gardens.

The clearance of the wildwood started in the Bronze Age, continued through the Roman occupation, and proceeded rapidly in Anglo-Saxon times. Again the woodland was being removed to feed an increasing population. So, by the time the Domesday Book was written in AD 1086, the woodland had been reduced to fifteen percent of the total land cover. It is estimated that it stood at thirty-three percent in the sixth century. So there was an amazing removal rate of thirty-two acres per day for five hundred years! This rate was not surpassed until the twentieth century and then we had chain saws and heavy machinery, not hand-axes and ox-carts for removing the felled timber.

Clearance was achieved not only by the axe, but by fire and animals, for the axe would not have been quick enough on its own. The fire would help to clear the land, which would be fertilized by the animals grazing on it after burning had finished. No new shoots, which would appear after burning, would survive the grazing animals, preventing woodland regeneration. This is the same slash-and-burn technique, which is reducing the South American rain forests today. As well as the much needed fields, some of the land cleared by the Anglo-Saxons was left open. There was probably too much to cultivate at any one time. Here our resilient flora found new lands to conquer and the heathlands appeared.

Lowland heaths are common in Britain, but curiously limited in distribution in mainland Europe. In Britain they are found on all types of soil, from sandy, free draining ones through to peats. Their main characteristic is that no one tree, grass, shrub or herbaceous plant is dominant. More homes for our frequently evicted flora, which survives being moved about, if competition for space is not too severe. If the grazing by deer, sheep or horses is not too heavy, rowan, silver birch—an Ice Age survivor—and aspen will appear. Shrubs include heather, which is not confined to the moors, several heaths, bilberry and at least two species of gorse. The common herbaceous plants are a mixture of heath bedstraw—one of the early rock garden plants—heath milkwort, and the delicate purple, parasitic lousewort in the damper places. Bare patches often have clumps of sheep's sorrel, which was an early alternative source of vinegar, hence its species name *R. acetosella*. These heaths were the fore-runners of the Norman Royal Chases and deer parks.

Since there was such a great loss of woodland, it is logical to assume that we have lost many species of plants. Unfortunately we cannot be certain, since the evidence is not clear enough. On a positive note, more hedgerows were created and footpaths and roads developed, as well as the new habitat, the heathland. All these events provided places for our re-invigorated flora to become established. This had been a dark age for accurate plant knowledge, which preceded an era of enlightenment.

6
1066 ETCETERA

Peace hath her victories.
John Milton

The defeat of the Saxon King Harold at Hastings in 1066 signalled the start of a new period of political stability in England, and the third phase of plant change shown on page five. The Normans established themselves on a land with a thriving population, based on intensive agriculture. The next 250 were medieval years of plenty. The climate was slowly becoming warmer. By the twelfth century it had reached a peak of warmth, which remained unchanged for over 200 years.

This warm climate allowed the renewed cultivation of the old Roman vines, creating vineyards at least as far north as Worcestershire. Their presence can still be seen in place names such as Vinesend. These vineyards were more productive than the French ones of the time. What would the E.U. say to that, and what quotas would be imposed? Unfortunately, we do not know anything about the quality of the wines. It would be nice to dream that Holborn, with vineyards recorded in the *Domesday Book* in 1086, produced an excellent vintage.

The open field system of the Anglo-Saxons, slowly modified by Norman feudal ideas, developed into the manorial system. For the plants this was ideal, for there was still room for the weeds and ruderals of the Anglo-Saxon period. Woodland boundaries became hedgerows, as the land behind them was cleared for farming. Some farmers planted elm in hedgerows for extra cattle fodder, and hazel, always a slow natural colonizer, was planted as an extra food source. Planting in hedgerows was a method of protecting the saplings as they grew, as exposure to the elements and grazing animals was reduced. It also left the fields untouched, available for grazing or cultivation.

These hedgerows became a safe haven for many woodland plants. Dog's mercury, wood anemone, bluebell, and primrose are found, indicators of early hedgerow sites, formed from woodland clearances several centuries ago. Many shrubs and trees which are slow colonizers, such as field maple and service tree, may not have survived into the twentieth century if there had been no hedgerow development. Their presence in a hedgerow is an indicator of antiquity, so the importance of these hedgerow refuges cannot be over-estimated.

The Normans developed deer parks which were usually woodland marked out on privately owned land. The boundaries were fences cut from locally available wood, built high enough to keep in the small fallow deer. Sources of less luxurious meat were also herded here, usually sheep, pigs, and cattle. The effects of these animals on the woodland is hard to imagine, for only pigs could find natural food in a woodland. Perhaps it was to prevent the animals from straying? These parks declined towards the end of the Middle Ages due to the loss of manpower after the Black Death. The effect on the flora can only be guessed; some plants would do well and survive, while others, unknown, would be lost.

The Normans introduced rabbits into England both for their flesh and their skins. Inevitably they escaped from their carefully constructed warrens, as anyone who has kept a rabbit will understand. In the medieval period, their effect was probably minimal, but today it is enormous. They nibble at new shoots on trees and other plants, turning, for example, heather heaths into grassy ones, which revert back when the rabbits are removed. The loss of rabbits due to myxomatosis revealed new plant communities in many places. These may well have been there before the rabbits crossed the channel from present day France. For us, it was a chance to look back in time, at some of our lost flowering plant communities.

The Royal Forests, the monarchs' supreme status symbols, were not fenced like deer parks. They extended over the heaths and moors created by the Anglo-Saxon woodland clearances as well as over established woodland. As long as the woodland was used for hunting, all was well for the plants. Unfortunately Royal poverty allowed the monarch to sell assarts, which were clearances of the Royal Forest, for agricultural purposes. The king could then claim a percentage of the crop, a modern rent or licence fee. It was not a good deal, for in addition there was a limit on the height of the boundary fences or hedges of four feet six inches, which meant that the Royal deer could get in, feed, and get out with impunity. So the possession of an assart was a mixed blessing.

On the positive side some of those Royal Forests survive such as Waltham Forest, Epping Forest, Savernake Forest, and many more. Although the range and number of plants has decreased in number, due to pressure of use, floral relics of these great forests are still here today. There is already evidence of regeneration in forests where the old management techniques are being used. Some of the old assart boundaries are still in place, and these are providing plant refuges for local flora, in the same way as the old Anglo-Saxon hedgerows. We are now more aware of the ecological needs of these great forests, and hopefully they will continue to thrive.

Records show that many of the planted Anglo-Saxon village boundaries have survived as modern parish boundaries, since they were unaltered for the most part by the Normans. These usually comprise a bank and ditch, with a hedge on top, representing the local floral diversity of the eleventh and twelfth centuries, relatively unchanged. In the twentieth century, they represent splendid pieces of floral history. Plants like service tree and field maple have been mentioned in this context; others include spindle, midland hawthorn, black bryony, the native wych elm, plum, cherry, crab apple, pear, blackthorn, and several species of wild rose. Hedgerows marking parish boundaries are often dated using Hooper's method, which relates the number of plant species present to the age of the hedgerow.

The common land not yet enclosed was still there for grazing cattle, pigs and sheep. One change or new development was the creation of gardens. This reflects the stability of life brought by the Normans, as the *Pax Romana* had brought ten centuries before. There was also a vast improvement in communications with Europe, opening a new cultural era. The monasteries of Europe had always had links with their English counterparts. These early, often tenuous connections, flourished with the improved contacts between religious houses. One of the results was an expansion in the exchange of plants in both directions. By the start of the thirteenth century, the monasteries had become established as trading centres for plants, acting like the nursery men of the nineteenth and twentieth centuries.

Man had to be well fed and have a relatively acceptable standard of living before he had the energy to think of plants for pleasure, and not just as a source of food, medicine, and other essential items in the household budget. This idea was already appearing in Europe, as early as the ninth century. The lists of plants in Charlemagne's book, *Capitulare de Villis*, include some which can only have been grown for their colour, since they have no known practical uses. Life must have been easier in Charlemagne's kingdoms than in England, for we did not reach this stage until the thirteenth century.

Although no medieval gardens survive intact, contemporary accounts describe walled or hedged gardens, with gravelled pathways around the Roman style, raised beds of flowers and herbs. These were usually sited so that they could be seen from the manor house or the castle's private apartments. As time passed, larger manor houses were built, so that as well as these herber gardens, there were ornamental moats and ponds. This produced large areas to be planted and landscaped from the monastery mail order system.

The reconstructed thirteenth century herber of Queen Eleanor at Winchester, grows plants whose identity is drawn from contemporary literature. In this garden, laid out in a medieval style, grow flowers such as daisy—oxeye species; native violets and the Roman periwinkle, with its climbing and sprawling growth habit; St John's Wort—often found on heaths and a plant of fairies and elves, and the flower of John the Baptist; pennyroyal—the Roman minty-flavoured herb; the tiny blue flowered speedwell; the cheeky-faced heartsease; sneezewort—which was a medieval cure for toothache, and its denizen cousin feverfew, the housewife's aspirin.

Feverfew is a medieval introduction from southern Europe, probably via the monastery mail order service. It is now a ruderal on waste ground, having escaped over the garden wall many centuries ago. Today it is still a medicinal herb, being for many people, a successful treatment for migraine. The turf, not a lawn as we know it, was a mixture of low growing, native, herbaceous plants. These include bird's foot trefoil, chamomile, cinquefoil, lady's bedstraw, selfheal, and ribwort plantain—another Ice Age survivor which plagues the gardeners today. There is nothing new in a garden!

In with all these native plants and a few denizens, the garden had a new and exotic plant. Queen Eleanor is credited with the introduction of the hollyhock, collected in the Holy Land. It grew well in her native Spain, so she brought it to England to grace her garden when she married Edward I. It is a quirk of fate that a Mediterranean plant should become the symbol of the English country garden, which like others before it, never left the garden to grow in the wild.

Modern planting in Eleanor's garden includes only those flowers which would have been familiar at the time, around 1272. The spring planting includes columbine, primrose, and cowslip, pansy and sweet violet, periwinkle, and wallflower. For the summer there are irises, peonies, bellflowers, borage, cornflower, and feverfew, hollyock and lavender, Madonna lily, and opium poppy, soapwort, and tansy. This collection represents a small floral history on its own, parts of which have already been described.

Gardens and gardening were becoming a part of the English way of life. By the middle of the thirteenth century, there were commercial seed centres in London and Oxford. One enterprising London nursery provided a ready-made garden, consisting of flowers, trees and turf. It is a marvellous flight of fancy to try and recreate that flower catalogue. Would it have included the native flowers from Queen Eleanor's garden, as well as those which arrived by the monastery mail order service, returning pilgrims, and of course Lady Luck.

The increased traffic between England and France inevitably brought new plants, travelling by accident in goods and on clothing as accidental hitch-hikers. The ballast and building stone from the quarries around Caen are the probable accidental transport medium for the yellow wallflower. The limestone was excellent for carving, since it was such an even-grained stone. The small seeds must have arrived in the cracks of the imported stone, and would have fallen off later, during transport to the building sites.

Caen stone was used in church buildings from Winchester to Canterbury and north to Durham City. No wonder that the plant spread. It still grows on rocks and cliffs throughout central, western and southern Europe as well as in Britain. Its real place of origin, many centuries ago, was the Aegean archipelago. The numerous hybrids of this plant are among the most popular bedding plants available here today, relics of the trade in stone. Another stone-loving plant arrived at this time, probably as a herbal remedy for pin worms. Bulbous corydalis—unfortunately sometimes also called purple fumitory—made an escape into the wild. Did it also arrive on stones like the wallflower? It is a possibility since its herbal use is limited.

There were other arrivals around this time including the creeping bellflower, a graceful flower but of no known use. A fellow arrival and present day rarity, the rampion harebell, was brought in as a winter salad vegetable. Perhaps plant identification then was not as accurate as it should have been. Bellflower, being very similar in appearance to its edible cousin, may have been introduced by mistake. Then there was a trio of plants which appear to be deliberate garden introductions. All have since escaped from the flower bed to join plants like feverfew to become vagabonds and denizens in the wild.

First of the trio is the very versatile tansy, probably a plant in the monastery mail order service. Its medical uses included the treatment of kidney inflamation and uterine conditions, especially abortions. A culinary use was in yellow Tansy cakes, a medieval Easter delicacy, as well as in the home, where it was the housewife's mouse repellent when placed in clothes presses. It came from southern Europe, and reached New England with the Pilgrim Fathers. Here it also left the gardens and has achieved native status, along with other garden escapes.

The second member of the trio is lungwort, which probably arrived in the same way as tansy. With its spotted leaves looking like lungs, it was inevitably used in the treatment of chest complaints. Today, it is sometimes found as a garden escape, though never far from the original flower bed. The spots on the leaves vary in colour, not a separate species but a reflection of the soil pH. The origin of its other common name, Joseph and Mary, is obscure.

The last member of the trio is garden angelica, introduced for its culinary and medicinal value. The leaf stalks were, and still are, preserved in sugar for confectionery, while the roots and fruits are used to make an effective tonic wine. Like the others in this trio, it made it over the garden wall to the land outside. Here it joins other members of the large parsley family, with their delicate, feathery, green leaves and irritating similarities, which make identification hard work. There is also a native angelica, and the two are easily confused.

By the end of the warm thirteenth century, everyone seemed to have a garden. The monasteries, which had always had land under cultivation, extended their coverage to develop individual vegetable gardens, plots of culinary and medicinal herbs, with separate orchards of fruit trees. The walls around these places often supported vines and the more tender fruit trees, such as apricots.

The working man, as well as cultivating an assart if he had the money to buy one, also had a backyard. This ranged from a few square metres to several acres in extent, and contained everything growing together. Cabbages and marsh marigolds—that Ice Age survivor, providing a yellow dye and a fertility symbol, for hanging over the cow byre door—lilies, lettuces, and leeks were all grown together alongside the bees and the chickens. Companion planting was definitely the method used to keep pests down, and is a popular method again today. The medieval gardener clearly knew something which has taken us seven centuries to rediscover.

The leek, brought here by the Romans, was a multi-purpose plant, providing a hair bleach—imagine the smell; toothpaste—it would certainly hide bad breath! and a catarrh cure, as well as being a useful vegetable. For better or for worse, it has not become naturalized in our flora, remaining just a very popular garden vegetable. As well as other vegetables such as onions, the average garden plot also included herbs like the native chickweed for the hens, and the two denizens borage for drinks and fennel for cooking purposes.

The fennel reflects the very poor quality of the meat, which was either very high or very salty. It would also serve as an cure for the resulting indigestion. The working man, and woman, would be like Chaucer's Sompnour who 'well loved he garlike, onions and lekes', for after all they were easy to grow and made the meat palatable. The use of garlic is curious for the garlic and onion are in the same family. Today garlic is a distinctive cooking herb, but the use of the name 'garlic' for other native onion-like species is common. Was Chaucer referring to one of these plants? We shall never be certain.

The pattern of farming now included an increasingly large volume of wool production. Something to clean the wool before it was dyed and spun would be an advantage. A southern European plant, used by the Arabs, Greeks and Romans called soapwort, was introduced. How it arrived is not clear, perhaps in a trading network? If a bunch of its flowers and leaves are shaken in water a lather is produced, which will degrease the wool ready for dyeing. Today, many shampoos contain soapwort to do just the same thing in a natural and harmless way.

The green or evergreen alkanet and dyer's madder also arrived around this time, possibly in association with soapwort, since they are both dye plants. The roots are sources of red dye, and the two plants were probably grown in the garden with the herbs. Inevitably some plants escaped to become established in the wild, often around farm buildings. Another arrival at this time was horseradish, an additional plant to hide the taste of tainted meat. This too escaped into the wild and now is a troublesome weed in wet places.

The Normans naturally brought symbols of their nationality and culture with them. The *Fleur de Louis*, named after Louis VIII, who chose it for his shield emblem, is known today as the yellow flag iris. It became known as the *Fleur de Lys*, and Edward III incorporated this iris into his coat of arms, all because the English could not disinguish between a lily and and iris—or was it how the name was pronounced? *Louis* and *Lys* are very similar to the ear. It is certainly not a lily shape. White and blue flag irises reached Italy from Syria around 1500 BC, arriving in England with the yellow flag iris. Like many other plants from the Mediterranean the white and blue irises have remained in the garden as gracious border plants. The yellow flag iris, apart from becoming established by rivers and streams throughout the British Isles, was a source of black dye and black ink from its rapidly growing rhizome system. This rapid growth and spread is an indication that it must have found the British environment very much to its liking.

The development of gardens for pleasure continued into the fourteenth century, when there was a change of direction. Friar Daniel started his own botanical garden in Stepney. He collected wild flowers such as bluebells, foxgloves, stinging nettles, teazles and mullein—did he call mullein 'Aaron's rod' , as some people do today? Records show that the practice of collecting wild flowers started in the century before, when many gardeners collected plants from the wild for their gardens. They would be easy to collect, and free, an important point at that time.

The most mentioned trees in this context were apple and hazel, to provide food as well as to give pleasure. Kingcups—also known as marsh marigolds, again a multi-purpose plant; the Welsh poppy; Jacob's ladder from the Great Rock Garden; the lenten lily, and snake's head fritillary complete a short list of the most commonly mentioned plants. Some of those gardens must have been truly splendid at the time and as we visualise them, looking back centuries later. We must not criticise medieval man, as we would one another today, for collecting plants from the wild and planting them in private gardens. Without his collection of plants in a garden, acting as a resevoir of seeds, many of these plants would not be here for us to enjoy today.

By the end of the thirteenth century the winds of change were blowing for the second time. All the land easily cultivated had been used but was still not sufficient to feed the ever-increasing population. The marginal lands were farmed to try and make ends meet. The fens were drained, again, as the Romans had done centuries before. It did not produce usable land then, and the same happened this time. The terraces of the Downlands were cultivated for the second time. These events affected the flora, particularly the herbaceous plants, which had enjoyed a relatively peaceful time since the Norman invasion.

The naturally occuring plants were pushed to the margins again, as man frantically tried to produce more food. With the reduction of the wildwood in Anglo-Saxon times, there was now a shortage of wood for fuel. The farmers burnt the straw usually used for manure, which further impoverished the soil, reducing its fertility, and lowering the crop yield. The other familiar factor was the slow cooling of the climate, an exact recreation of the events at the end of the Bronze Age. As shown in the Ages of Plant History on page 5, the Second Biological Crisis was taking place, due to an increasing population on an over-exploited land.

If this was not a sufficiently large disaster, there was worse to come. The Great Plague or Black Death first arrived in the summer of 1348, killing one third of the population within one generation. It is probable that the black rats, native to the eastern Mediterranean and with their plague-carrying fleas, arrived here on the returning Crusader ships. The Black Death left numerous villages completely abandoned, allowing the open fields of arable land to become overgrown in the absence of any workforce. Agriculture has never completely subdued nature; it suppresses natural succession but does not prevent it. Back from the margins of the fields and woodlands came the plants to reclaim their own territory, just as they had done when the Romans left over nine hundred years before. In the face of another natural disaster, there was something good happening.

Norfolk's Breckland is a good example of a marginal area pressed into agricultural service when the thirteenth century population got too large in that region. Even so, there was only a sparse population by modern standards, probably only twenty-five people per square mile, compared to the present East Anglian average of seventy to eighty per square mile. The land had been developed for sheep farming using the large areas of open land which had been created in Neolithic times. In addition there were numerous large rabbit warrens, set up in the eleventh or twelfth centuries.

The Black Death caused so many deaths that twenty-eight villages were left deserted. The plants returned from the edges, covering the untended meadows again, since there was no one to farm the land. Hawthorn would have been the first shrub to appear. It withstands browsing animals, especially rabbits and sheep which would have been in abundance and unchecked. In this shelter other plants would become established, eventually moving out to recolonize the land, sometimes to the present day. Many deserted farms and villages still lie under a cover of regenerated grassland and heath. Perhaps they should remain there in peace? Some have been ploughed over, some have been built over, and others fought over. The last battle of the Wars of the Roses, at Bosworth Field in the Midlands, was fought over the deserted village of Ambion.

Although the flowers and shrubs returned to the Breckland, the woodland cut down by Neolithic man did not. That is gone for ever. Now there is a unique heathland flora, developed after both Neolithic man and the Normans had left. Today the recovered land, with its unique heaths, is being destroyed again, this time with air bases and their associated activities. Do we never learn? The same story is repeated in the Weald, where the Normans felled over four hundred and fifty acres of woodland. The medieval farmers did not develop any large scale woodland management techniques; the early coppicing skills seemed to have been lost. Curiously, the Romans managed this area well, keeping their foundries supplied without destroying their raw material resource. Woodland never returned to the Weald and today there are Downs, large areas of grassland, where there were once forests of hornbeam, oak, and beech.

Many historians feel that the move towards sheep farming was not the result of the Black Death, but that it came later in the fifteenth century. Be that as it may, there is no doubt that the cultivation of the marginal lands ceased. The labour force was so reduced that the landowners started to compete for tenants. In East Anglia the number of sheep had been rising before the Black Death, with Norwich becoming a large centre for the woollen cloth called worsted. This had been introduced by the Flemish weavers who had settled in the area. It meant that the plants could recolonize the old arable land while new communities developed on the grazing land. For a while the rest of the woodland was safe.

The good grazing land and readily available supplies of water made other parts of England good sheep rearing country. Shropshire,

Herefordshire, the Cotswolds, and the Mendips all became inportant centres of the wool trade. The Roman town of Cirencester became the centre of the medieval wool industry in the west of England, due to its position on the north-south crossroads of Roman origin. Other centres included Chipping Campden, Chipping Norton, and Shrewsbury. In all these areas, the woodlands were depleted even further to allow for the grazing sheep, whose nibbling eating habits prevented any regeneration. History often repeats itself, for when sheep farming declined in the early stages of the Industrial Revolution, the trees, shrubs and flowers did not return to the fields. The damage to the land had been too great.

Some of these uplands are still covered with grass, for there has been no recolonization by the the original flora. Others have become areas of arable and mixed farming. The Cotswolds, for example, are covered with extensive fields of cereals and yellow oil seed rape. Like the moorlands created at the end of the Bronze Age, we have to accept and appreciate this smooth landscape, and enjoy to the full its unexpected richness. The valleys, protected by their unsuitability for grazing, contain some of the flowers which must have bloomed there before the sheep came. The spring plants which thrive are bluebells, primroses and enchanters nightshade, while the streams have water plants such as marsh marigolds—that great survivor—and water crowfoot. In the summer there are numerous orchids and potentillas, along with flowers of the chalklands such as sainfoin and scabious. In the hedgerows there are the native species like service tree and field maple, with elm and spindle surviving the agricultural activities of man in a safe refuge.

As this phase in our floral history closes, it is possible to look back at the changes which have taken place. The more vigorous native plants have survived the numerous invasions, advancing and retreating across the land as agricultural practices have changed. New plants have arrived, some by design, others by chance, which have naturalized in our flora. They have formed a second strand in the evolution of our floral history. The next period will add considerably to our knowledge of the native flora, as well as record the numerous, invading hitchhikers and other newcomers.

7
AS WE LIKE IT

For the glory of the garden it abideth not in words.
Rudyard Kipling

Still in the third phase of plant change, the period from the Black Death to the Industrial Revolution spans three royal dynasties and two civil wars. Within this turbulent period of our social history, there were periods of great stability. More people became literate, and there was a rebirth of accurate drawing of plants and animals not seen since Roman times. Travel began to extend beyond Europe, as the population became more forward looking and inquisitive.

The first dynasty was the Plantagenets, whose name is plant based on a corruption of *planta*, a sprig, and *genista*, the broom. This dynasty was feudal and insular. The Wars of the Roses had left the country weak and in a profound economic decline. The wild plants, back from the margins after the Black Death, maintained their places and consolidated their positions, ready for the next onslaught on their territory, which came with the Tudors.

This royal house was very different from the Plantagenets. They were forward looking and thirsty for knowledge. They moved England slowly towards the Industrial Revolution as a result of their lifestyle. Castles went out of fashion and houses were built with proper gardens and parks around them. They had glass in their windows, which were set in wooden frames. There were iron grates for the fires, iron cooking pots, and iron farm machinery. Weapons like cannons came into use in a big way. The navy had started to expand under Henry VIII—remember the Mary Rose?—as did the commercial fleet, to establish trade with the Mediterranean and later on with the New World.

It was Elizabeth I, acting as a royal catalyst, who finally revitalised the navy. 'This scepter'd isle' was under siege from Spain so we built our ships 'against infection and the hand of war', Think of the size of today's warships and their weight—a cross channel ferry is some 29,000 tonnes—and now realise that the Elizabethan navy successfully fought off the Spanish Armada with ships ranging from 50 tonnes to 400 tonnes. The oak for these ships was cut from the woodlands which had survived the Norman onslaught. Later on in this age of expansion, this use of timber was to become a problem.

During this second dynasty there was also the expansion of literacy and art. It was a period of peace, despite the Spanish Armada, so that people had the time and inclination to learn how to read and write. They learned how to draw accurately, not making their subjects into legendary beasts and grotesque plants. So at last there was a literate and scientific approach to recording our flora. Lists of plants grown in gardens were written, their uses for medicines and about the house were recorded, and details of their origins, both home and away, were listed. There was correspondence with gardeners in Europe to add to the information about plants at that time. This new era of learning in the Tudor and following Stuart dynasties would add several more coloured threads to the strands of our plant history.

Sixteen centuries of life on this island had to have affected the plants which were known to have been here before the Romans came. Pollen records become increasingly sparse as time approaches the present. By the fifteenth century, there is little evidence of value. This age of writers began to fill the gaps. There were no handy floras to help with the identification of flowers, so the early writers made up new names, or used the familiar local ones which varied from village to village. These names sometimes described the shape of the flowers and often the places where they were found. Kingcup, sundew, and Lady's tresses are all names given to plants which describe how they looked at that time. One writer John Gerard, called wild clematis 'traveller's joy' and *Viburnum lantana* the 'wayfaring tree', because both of them were found by the roads or ways. These very descriptive names from the sixteenth century are in use today acting as reminders of one man.

A Northumberland doctor and apothecary William Turner, c.1508–68, was the first person to describe accurately and list up to three hundred species of plants in his native area, including bluebells, dog's mercury and yellow flag iris. The name primrose, so easy to us today, came from the medieval Latin *Prima rosa*, the first rose of the year. But did the early literature mean *Primula vulgaris* or *Primula veris*? Turner had trouble separating 'primerose' and a cowslip or 'cowslap ' or 'pa(i)gle'. Parkinson nearly eighty years later writes with the same confusion, or perhaps there was another species which is now lost to us.

Turner was physician to the Duke of Somerset at Syon House around 1548, the time in which he invented the name cow parsnip, today's hogweed, as 'English cow's persnep'. The spelling has changed a little between the sixteenth century and today, highlighting a problem which we now face when interpreting these early records. Who would read 'pagle' as cowslip, definitely a local name for this pretty flower? Turner's *Herbal*, Part One, was published in 1551, being the very first account of the naturally occurring English flowers which were the doctor's only weapon against disease at that time. He completed Part Two of the *Herbal* in the reign of Elizabeth I, to whom he dedicated his work. The gap in time was inevitable, because, as an outspoken Protestant, he had spent the reign of Mary Tudor in exile to avoid being burned at the stake.

Turner mentions two plants which have made a successful bid for freedom over the garden wall. The first is melilot which he records seeing in its native Friesland. It is probable that he brought it back with him, since it is listed in the Syon House garden accounts for that time. It was used in a melilot plaister, or plaster, and as a strewing herb. Do we owe this tall graceful plant of waste land and verges to the Father of Botany? —a nice thought.

The other plant he records as a successful garden escape is tamarix or tamarisk, a plant of south eastern Europe, north Africa and the Canaries. It was growing in Germany when he records it around 1548. Later on Edmund Grindal saw it as well, for he was also in exile in Germany, escaping the fires of Mary Tudor. When he returned as Bishop of London, he planted a tamarix at Fulham Palace. Like melilot it escaped, and now lives by the sea along the south coast and on the Isles of Scilly. Here it withstands the gales and the salt, just as in its ancestral homeland; a plant with Protestant as well as garden connections.

In this age of experimentation the Elizabethans began to grow new vegetables, such as savoys from Holland, New World potatoes and tomatoes, rice, and runner beans. These plants never made a bid for freedom, but the exposed earth was a new place for the plants outside the garden. So in came what the Elizabethan gardeners doubtlessly called weeds. How many were able to find permanent refuge here we shall never know, but there are weeds of the garden today which never leave. Did they arrive here in the Elizabethan era of proper vegetable gardens?

Another source of plant information comes from the writing of Thomas Tusser. A gardener of great stature, his accounts of husbandry, in verse too, give a glimpse of agriculture month by month in the sixteenth century. In his advice on how to gather mustard, manage the pigs and the corn meadow, willows, 'wynnes'(?), and furze for faggot hedging, we see a knowledgeable farmer who knew his plants. His spelling varies, highlighting the problems of plant identification for us in the twentieth century.

Tusser refers to 'cowslepes' and 'paigles, 'gellyflowers', 'stock gillyflowers' and the Queens 'gilliflowers' among his herbs and flowers. It is not always clear why some of these plants were grown; they must have had a use. There is an impression that the climate was warmer, judging by the volumes of crops to which Tusser refers. One of the most amusing garden hints was to 'grow strawberries under gooseberry or barberry'. This would certainly protect the fruit from the birds, but picking it would become a painful experience. Perhaps the bushes were a different shape in the sixteenth century, or the prickles less aggressive?

Amid all this excitement of the identification of plants old and new, warning bells were ringing. These new houses and gardens were on land that had been cleared from woodland. The Elizabethans, a forward-looking people, were becoming aware that the apparently inexhaustible supply of trees was running low. It was not only the land for houses and gardens which was removing the woodland at an alarming rate. Coal was replacing some of the demand for energy, but it was mined by the open cast method which needed the removal of the woodland first. The wood for charcoal burning was removing the woodland in the Weald, which had suffered since before the Romans. Our native greenwood tree, the common or pedunculate oak was needed to build ships for our 'gold and glory boys', Drake and Raleigh, who were protecting us from the wicked Spanish with a spot of profitable piracy in-between times. At home we needed wood for the domestic market and house building; supply was not equal to demand.

'Plant a tree' campaigns are nothing new. Lord Burghley, Queen Elizabeth's chancellor, foresaw the navy's problem quite simply as 'no oaks, no ships'. He promoted the planting of acorns interspersed with holly as a protective crop. Protection from what, animals or man? There is no record of an answer to that question. Military necessity had affected the size of our gardens, for these oaks need a wide space to grow. The wide, sharp angled branches were needed for ship building because they could support cannons in the heavy rolling seas of the

Atlantic. Now we had a beleaguered army of plants whose territory was lost as the woodland was cleared. As they had done before, they returned to colonize the new places in the tree nurseries of Tudor England.

The Tudor Rose of learning which had started to open at the start of the sixteenth century now came into full bloom. Following in the steps of Turner, another apothecary, physician, and master surgeon made a huge, if somewhat controversial contribution to our plant knowledge in the late Tudor period. John Gerard, 1545–1612, had a physic garden in Holborn and from it in 1596 produced a garden catalogue listing a thousand or more herbs and rare plants which he had grown. It is a permanent record of some of the much-loved herbs and plants which we grow today.

It is probable that William Shakespeare, who lived nearby, sat in this oasis of sweet smelling plants to write many of his plays, rich in the old English herb lore learned from his friend John Gerard. There are references to a bluebell in Cymbeline calling it the 'Azur'd harebell', which begs the question bluebell or harebell? They are two different flowers. Again the problem of a change in word usage. King Lear refers to samphire which Gerard grew, and Ariel mentions wild thyme, another plant in Gerard's garden. Being country born, did Shakespeare just use the plants he remembered from his youth? We shall never know. This is another coloured thread in the strands of our plant history for we learn of plants through another medium.

Gerard's *Herball* of 1597 is a masterpiece of glorious English and folkloric comment, as fresh today as the day it was written. Mention has already been made of the names which he gave to plants, but there were others he recorded, which were the old English ones. Jackanapes on horseback—double marigold—and 'go to bed at noon'—goatsbeard—are but two of from the long list. He describes the uses of plants in an amusing way, for example the use of Solomon's seal as a remover of bruises 'gotten by fals, or women's wilfulnes in stumbling on their hastie husband's fists'.

John Gerard was one of the first gardeners to grow potatoes, around 1586. It is possible that Francis Drake had collected them from Virginia along with other plants, and given them to his friend. However, many historians feel that Drake found them in a ship's stores, when raiding on the Spanish Main and gave some to both Gerard and to Raleigh. The latter is usually credited with being the first to grow potatoes in England and Ireland.

It is known from letters that Gerard had other friends who sent him plants. Men like Thomas Hesketh who sent him wild rosemary from his native Cheshire; a country doctor Stephen Breadwell along with a fellow apothecary and tulip fancier James Garret, sent him herbs of interest. He had a high political profile with friends in the city and abroad. Nicholas Lete, a city merchant sent him 'gillofloure with yellow floures'—wallflower?—from Poland; Lord Zouch sent seeds from Spain, Italy and Crete. Another regular correspondent was Jean Robin of the Royal Gardens in Paris, who sent among many plants, the local barrenwort. Unfortunately there are no records telling whether the gifts grew, though barrenwort certainly did. It is found today in damp and shady places in the north of England.

The *Grete Herball* represents hours of patient exploration around what we know today as London, a huge area covered in cement and tarmacadam. At the end of the sixteenth century, there was a City of London and the areas round about it were separate villages. The *Herball* records mallow outside the City by the gallows at Tyburn—now near Marble Arch—mullein near Highgate village, and clary in the fields near his home in Holborn. North London, now part of the great urban sprawl, was covered with woods and fields, in which Gerard records yellow archangel, twayblade and the rare bird's nest orchid. Almost five hundred species of wild plants are listed of which one hundred and eighty two are new recordings.

Even with the undoubted plagiarism and all the emotions which that brings, the *Herball* is a magnificent piece of work, by a man whose love of flowers and their history shines through every page. It is a marvellous account of our flora at the end of the sixteenth century, as well as giving great pleasure to those who read it today.

The gardens of the increasingly wealthy middle class, as well as those of the nobility, had new and unfamiliar plants. In this new age of travel, plant hunters had started to bring home an exciting range of trees, shrubs and herbaceous plants from abroad. Laburnum with its long yellow tassels and dark timber, a cheap substitute for ebony, had arrived from south east Europe, the horsechestnut with its familiar candles and conkers arrived from Albania. Buying trips to Italy and Bulgaria produced the yellow winter aconite and the regal martagon lily. Gerard records a gift of martagon lily from Garret but does not say where it came from, nor does he record the gift as bulbs or seeds. Unfortunately these well-known trees and plants do not naturalize and have remained safely in the garden. Oliver Rackham drily puts it that the horsechestnut has become the 'universal tree of bus stations'. It must be concluded that this beautiful tree is immune to diesel fumes!

Three plants arriving at this time and eventually becoming naturalized into our flora are sweet flag, soldier's pride or red valerian, and thorn apple. Gerard names sweet flag in his garden list of 1596 calling it a relatively new plant and dating its introduction from Turkey around 1567. It was a plant of the Indian bazaars as well as a strewing herb in European cathedrals, due to its pleasant smell when bruised. Very useful at a time when personal hygiene was not very good. It spread out from the garden, rapidly colonizing the waterways and canals. Gerard also grew red valerian or soldier's pride, though only as a curiosity for he found no use for it. Today this calcicole plant from Portugal and the Mediterranean is found as a splendid urban colonizer on the mortar of walls, apparently immune to vehicle fumes. It grows well on the limestone cliffs of the Isle of Portland, where it is known as 'convict grass', a reference to the prison on the island.

Thorn apple is probably a native of eastern Asia. The small light seeds had been carried to Europe from Asia by the gipsies, where it became the Turkish alternative to opium. Gerard had received seeds from both Lord Zouch and Jean Robin. It had been found to be too poisonous for internal use and was restricted to use in poultices. In the days before laboratory testing, one wonders how the apothecaries found out that thorn apple was so poisonous—animal testing? Thorn apple reached America probably on ballast stones—remember the wallflower?—where it is naturalized in the local flora.

Did Gerard unwittingly record golden rod, *Solidago canadensis*, when commenting on a golden rod brought in from overseas? The native plant *Solidago vigaurea* was an excellent though expensive treatment for stab wounds, a common feature of life in Tudor London. When the new overseas plant was seen growing in Hampstead Wood and near the 'village of Kentish Towne', the price fell rapidly. Luckily for the stab victims the medicinal effects were the same. Today golden rod, *S. canadensis*, is a popular garden plant whose numerous seeds escape over the garden wall as they must have done in the sixteenth century. It lives in the wild with the native golden rod from which it is not always easy to separate.

How and why the blue and purple flowered greater periwinkle arrived at this time is not clear. It too is an infrequent garden escape, like its relation the lesser periwinkle which arrived with the Romans. Another garden plant arriving at this time was honesty, with its attractive silver fruits. It comes from southern Europe and Italy where it grows wild. It, too, is a rare garden escape although Gerard records it in the wild around Pinner, Harrow, and Hornchurch around 1570. Today it is often found in the hedgerows around old farm cottages, where the flowers can be white as well as magenta. Leopardsbane, which arrived at the same time as greater periwinkle and honesty, escaped out into the wild to habitats such as woodlands and damp shady places. It had a dual function, for as well as being a decorative plant it was used in the treatment of eye complaints.

It sometimes seems that the list of Tudor garden escapes is endless, for this age of gardening and increased travel certainly left its mark on our flora. Sadly, some of these escapes have almost disappeared. Cornflower, probably like melilot, came again as a Mediterranean herb for the physic garden and by accident, in the uncleaned grain imported to feed the hungry English. Gerard grew it for an eyewash and ink dye and describes it 'growing in corn fielde among the wheat, rie, barley and other graine.' Modern pesticides and grain cleaning techniques have almost eliminated it from our flora.

At the start of the sixteenth century the population was between two and a half million and three million people. In spite of the turmoil of Mary Tudor's reign, the lack of wars and epidemics produced the inevitable effect of an increased population. This meant that there had to be an increase in food production The second agricultural revolution had started, making the old land more productive rather than finding more space on which to farm. There was a more scientific approach to farming in line with the rest of our cultural advances. We copied the Dutch farmers, who had started to alternate wheat with white clover and grass seed as a fodder crop, which gave a better wheat yield the following year.

Clover was imported from the Dutch but it was not cleaned, so like the grain of the Neolithic farmers and the Beaker people, it contained seeds of other plants. In this way the common ribbed melilot and the small melilot probably arrived. Just to add to the confusion the common melilot was imported as a fodder crop, so it is reasonable to assume that it had two entry routes to our flora. Gerard records seeing melilot both as a fodder crop and as a weed in a neighbouring corn field. In the end it does not matter, for it has become naturalized here along with the white or Dutch clover, which also made a successful bid for freedom. It now grows and flowers freely on the the disturbed soils around cultivated land and footpaths.

A curiosity from this era of new scientific farming is sainfoin. Did we just grow native sainfoin or did we have to import it, because we did not have sufficient native seed? Gerard records it by its beautifully descriptive, old English name 'cockshead', as flowering in Bedfordshire and Cambridgeshire. Today it can be found all over the chalky hills of Dorset, Hampshire and Gloucestershire. Is its present day French name —meaning wholesome hay—an indicator that we imported it because the home stock was insufficient? If so, this means that the present day sainfoins are a mixture of native stock and denizens; but does that really matter?

Finally, the fringed water-lily with its bright yellow petals arrived from central and southern Europe. Today it thrives on slow-moving fresh water rivers along with the native plants.

Sainfoin

8
THE GREAT PLANT WRITERS

Many suffer from the incurable disease of writing.
Juvenal

The name of a Flemish writer appears in Gerard's notes on the uses of leopardsbane. Mathias l'Obel, a doctor and enthusiastic botanist, as many doctors were out of necessity, settled in England around 1561. He did some intensive plant recording around Bristol, finding around eighty plants which had not been recorded before, even by Turner with whom he was contemporary. L'Obel, whose name is commemorated in the plant name lobelia, became botanist to James I, so bridging the gap between the Tudors and the Stuarts. This transfer of dynastic power did not slow down the influx of new plants, nor the enthusiasm for botanical discovery, for the Stuarts were as forward-looking as the Tudors. If anything the pace quickened as the levels of literacy and financial support increased. At the head of the support queue was King James himself, who needed the services of the botanist l'Obel and later the apothecary Parkinson, for he was a very sickly man needing constant treatment.

Gardening was moving into a new phase, the prosperity of the Elizabethan era having produced a rash of great houses with large gardens. Robert Cecil, Lord Salisbury, keen to make his garden at Hatfield House one of the best, hired a young gardener-botanist named John Tradescant. The introduction of new plants was still slow and a little haphazard, and John Tradescant was hired as plant hunter, to look abroad for interesting garden plants for Robert Cecil. Some of his more colourful finds which included lilacs, narcissi, and cistus never escaped into the wild.

It may have been projects like this which inspired another form of gardening, more sympathetic to our flora. In 1617 the Society of Apothecaries was formed and with it, designs for a new kind of garden. In 1621 the Oxford Botanic Garden was opened by Henry Danvers, who spent £5,000—equivalent to five million pounds today—on providing a place to be planted with 'Divers simples for the advancement of the faculty of medicine'. Quite a change from Gerard's physic garden on land leased from Lord Burleigh. Our beloved herbs were to be safe, either in the individual gardens of a recognised professional,

5. *An old salt route over the Malvern Hills*

6. Modern Hybrid Hollyhock

7. *Houseleek*

8. The Botanic Gardens at Oxford

The Great Plant Writers

the apothecary, or in the botanic gardens of the larger cities. The latter are a far cry from the garden of Friar Daniel in Stepney, but they represented a step forward in the care of, and interest in plants.

It was at this time that more of the strands in our floral history began to come together. The work which had begun in the sixteenth century to record the native plants was being continued, but with a difference. The plants being recorded were not just the medicinal ones, but all the wild flowers. Thomas Johnson was not only an apothecary but a keen botanist. He organised the first recorded botany field trip, a four day walk-about in Kent with a group of fellow doctors, who listed all the plants which they saw. Their findings were published as *A Kentish Journey* and was followed by two more publications from similar field outings. The aim had been to produce the first *English Flora*, but this was cut short when Johnson was killed during the Civil War in 1664.

He recorded the rare lady's slipper orchid — note the descriptive old English name — in the north of England when out with John Tradescant, who was by then gardener to Charles I. Tradescant is credited with the introduction of the North American Michaelmas daisy, which Johnson records as growing in the countryside. Today it is a weed of waste land around railway stations and signal boxes; sites which were not around in Johnson's time. He also lists the evening primrose as an import from North America, much earlier than John Tradescant, Jnr., son of Cecil's man, who is usually credited with its introduction, could have brought it. It was certainly naturalized by the 1630s. Were the lovely long, yellow spikes of flowers, opening in the evening to give a rich perfume, the only reason for its introduction?

Johnson's work was not only an account of all the plants in the areas which he visited. He started to query, in a serious way, the status of some of the plants which he recorded. Were the rare, elegant star of Bethlehem and the common pale lilac vervain native plants? Gerard had hinted that he did not think they had native status and nor did Johnson, though neither suggested why this was so. Without a flora to help him Johnson, in his determination to be accurate, had considerable problems with identification. This resulted in recording eighteen species of scabious but admitting defeat with a range of hawkweeds. Today's botanists have a fellow sympathy with him, for the latter group is difficult to separate even with a modern, up-to-date flora.

Johnson's friend John Goodyear, a field botanist of the same high class, first recorded ivy-leaved toadflax in the Ockendon garden of a keen amateur gardener, William Coys. He was given some seeds which he grew in his garden in 1618 in Droxford, Hampshire. He cannot have been the only one to have had these seeds of ivy-leaved toadflax, for by the 1640 it had been recorded by Parkinson at Hatfield House as well as in other gardens in the area. The brick walls around gardens, parks and land boundaries, provided by the building boom of the next few centuries, gave ivy-leaved toadflax a huge habitat to colonize. It is one it has never lost, so today this Mediterranean plant with its ivy-shaped leaves and tiny purple and yellow flowers has colonized boundaries made of all materials except wood. Will that come next?

John Parkinson succeeded l'Obel as apothecary to James I and Charles I. It is always fascinating to speculate how much he might have seen of Tradescant and Johnson. Did these great botanists of the same era exchange information? In his writings Parkinson refers, like Gerard, to Nicholas Lete as a supplier of plants or seeds, especially from the eastern Mediterranean and Turkey. Parkinson's interest was also commercial since he had a garden shop. His first book, published in the same year as *A Kentish Journey* was really a catalogue of garden flowers.

Neither this nor his herbal of 1640 told us anything new about our flora. It was more a reinforcement of what had already been recorded. However he was a competent botanist, carefully recording the Roman nettle from a site near 'Lidde by Romney'. Here we see a problem with site identification, for the names of places have often changed their spelling through time, though this one is easy to interpret. Others are often more difficult to deduce. The places may have disappeared or been included in other, more recent settlements. Did Parkinson learn from Johnson, Tradescant, and Goodyear the precise site of the Lady's slipper orchid? It is not a conspicuous flower and he records it in Lancashire. Was this the 'North of England' of the others' records? Plagiarism is not new.

Parkinson travelled a great deal, including journeys across to Ireland. Here he records seeing the strawberry tree which is only naturally occurring in the south west of that country. It never reached England because the sea rose too quickly at the end of the Ice Age, cutting off Ireland from the rest of mainland Britain. Did he bring back cuttings or seeds of the strawberry tree for his customers, for it is frequently grown in parks and gardens here on the mainland?

It is his gentle comments about their uses which go a long way to explain why some introduced plants have survived and others have declined or been lost altogether. Pudding grass or pennyroyal, he explains, was too hot for the seventeenth century taste. Remember that the Romans who introduced it suffered from a permanent loss of taste due to their use of lead cooking pots. The milder spearmint was preferred for cooking and for medicine. What Parkinson did not know was that the Romans introduced spearmint as well, and that it was a Mediterranean plant. Turner and Gerard sing its praises and add the fact that among many other medicinal and domestic uses, it was an excellent strewing herb. All this accounts for its wide distribution today. The distribution includes North America, for the Pilgrim Fathers took it along with the other plants which they needed to start their new life. It made its escape into the wild, but is less common than here in Europe and England.

In complete contrast the pink *Dianthus plumarius*, a calcicole from the hotter areas of south west Europe, arrived here, whether by accident or design is uncertain. This pretty, frilly-petalled flower escaped into the wild from the garden, as well as being the basis for the breeding of garden pinks, when crossed with our native Cheddar pink. Perhaps the famous plant breeder Master Ralph Tuggie, lovingly referred to by both Gerard and Parkinson, used this plant to raise his new, sweet smelling species of gillyflowers, pinks, and sweet williams.

Here we have a new twist in the strand of our floral history, a wild alien and a wild, native flowering plant, together producing a new garden species. When the Roman carnation was added to the cross, the strand becomes even more twisted. Some of the larger, more exotic carnations on sale today have a history second to none. Next time that you buy these large carnations, you can look at them with renewed respect for their unique history and lineage.

Much better known and far less scholarly than Parkinson's herbal is Culpepper's *English Physitian* of 1653. This avowed republican naturally had no patronage, and having completed four of the seven years as an apothecary's apprentice, declared himself a doctor and published his book in English for all to read. As with Parkinson we learn little more about our flora, though his dry humour is to be appreciated. He remarks on eyebright 'If this herb was as much used as it is neglected, it would spoil half the spectacle-maker's trade'.

It is hard to imagine that in spite of all this work there still was no Flora of England. This was at a time when there was a newly-formed Royal Society and a renewed interest in scientific matters. John Ray was born in Braintree, the son of a blacksmith, and an 'Essex Man' of considerable academic stature. He published in 1660 a catalogue of the plants around Cambridge, where he was educated. The works of Turner, Gerard, Parkinson, Johnson *et al.* were all shaken up together and standardised, especially the names. He produced for the first time some order in our flora which all could use. Well nearly all, for he wrote his *Cambridge Catalogue* in Latin. It was a pity that he did not follow Culpepper in this respect. There were two more publications in Latin in 1676 and 1690, the latter being the first descriptive flora to be published. Ray was also the first botanist to separate the monocotyledons from the dicotyledons, something we take for granted today.

Try and imagine Ray and his constant companion Willerby, riding the lanes and byeways, looking from left to right from saddle height at the plants growing along the verges. What would make them dismount and look in a field as we might today—instinct? It makes one realise what a labour of love, along with a fascination for flowers, produced these early floras. How did Ray see and so record for the first time, opposite-leaved golden saxifrage and the rare alpine bartsia—was it rare then?—near Orton in the Lake District? He also recorded the northern bedstraw in this area. Ray added approximately two hundred new species to his lists, which with the work of others, brought the known total at that time to nine hundred.

A plant common today on railway sidings and pavements—a popular place for alien plants since there is less competition—is Canadian fleabane. Ray in his 1690 publication, lists it as a new weed near London, but which had been recorded earlier by one Tancred Robinson, profession unknown. The story is told that it arrived here in the stuffing of a bird, possibly a parrot. It was recorded in the Botanic Gardens in Blois in 1653, from where it had naturalized in the Midi by 1674. It remains in our flora as a plant of no great beauty, merely an historical curiosity.

While Ray was collecting plants for his flora, John Evelyn was recommending herbs in the diet as a prevention against disease, rather than as a taste enhancer. This was the first time that the plants had not

The Great Plant Writers

been listed for their medicinal properties. This salad-loving man was an authority on trees and recommended a tree planting programme. He was following Lord Burghley, who had a century before urged the planting of the pedunculate oak, for just the same reasons. When the Chelsea Physic Garden opened in 1673, John Evelyn visited it and recorded ivy-leaved toadflax growing on the walls, an intrepid colonizer then as now.

Two more notable plants appeared at the end of this century, Oxford ragwort and the Duke of Argyll's tea plant. The bright yellow flowered Oxford ragwort from the slopes of Mount Etna in Sicily was planted in the Oxford Botanic Garden, which had become a well-known collectors' garden. A hundred years later, it was recorded outside the garden around Oxford. The development of the Great Western Railway by Brunel in the next century spread the feather-light seeds even further on the wind currents created by the trains. The bright yellow flowers can be seen today, growing in the clinker of tracks and sidings around the country. It went across the sea to Ireland over a hundred years ago, managing to reach the Emerald Isle without the car ferries and is now spreading slowly across the country; a truly adaptable plant.

The Duke's Tea Plant

The Duke of Argyll had an interest in exotic plants and ordered for his garden in Whitten, Middlesex, a wide range of trees including a tea shrub *thea* from China. History does not relate which mail order company was used, but as often happens even today with our computer-based systems, a mistake was made. He received a *lycium* labelled *thea* and a *thea* labelled *lycium*. It is the trailing shrub *lycium*, the former, which we know today by the ironic name of the Duke of Argyll's tea plant, with purple flowers in olive green foliage and scarlet autumn berries.

By the turn of the century the plant had made a successful bid for freedom from the garden in Whitten. It is found today in hedgerows as far north as the Midlands, growing with adventives, denizens, and natives, a mail order mistake for all to see. The Duke would have had an unexpected gastronomic surprise had he made a brew from the wrong one!

Inspite of all the wealth of new knowledge, we were still nowhere near knowing or recording all of our flora. If today there are approximately two thousand species, how many were present at the end of the seveteenth century? Successive herbals tended to reinforce what was already known and the century closed with another excellent herbal from John Pechy, and the opening of the Botanic Garden in Edinburgh. Then the rumblings of floral distress and botanic warning first heard in Elizabeth I's reign began to get louder.

People had no taste for scenery at that time; it was a culture form yet to arrive. Great writers like Chaucer and Shakespeare had disregarded it, taking the wild places for granted. So when the rumbling machinery of the Industrial Revolution began to remove the wild places, taking moorlands for mills and woodlands for factories, and blocking and diverting streams for power, the population was unprepared. The plants, unless safe in the physic or domestic gardens of various sizes, were under siege yet again. This siege was one the flora was bound to lose because there was no one to fight its cause until after 1945. We had lived with our flora and used it throughout the centuries. Throughout time, from the Celtic farmers to the Roman introductions, from the Saxon herb gatherers to the gardening revolution, and finally arriving at the end of the seventeenth century with its plant records and early scientific farming techniques, there were plants to be used. What came next?

By the late seventeenth century the Industrial Revolution had produced methods of refining lead and arsenic compounds with medicinal properties. Culpepper's *English Physitian*, in English not Latin, had been a failed attempt to halt this trend to new and expensive, artificial medicines. The dilemma of the new urban population was one which actually saved the plants with a medicinal use. Working people could not afford the expensive new drugs based on heavy metals; the cheaper or inexpensive medicines available were still based on the old herbal remedies. This was particularly true in the industrial north of England. Of course in the rural areas, it was 'as you were' and not 'back to basics' because the new medicines had never been a serious option.

These rural areas responded and supplied the urban areas with the plants which were needed. This ensured that out of need, at least the medicinal flora would survive the holocaust which was to come. This fact alone saved a whole army of plants from the octopus-like death grip of the Industrial Revolution. By the nineteenth century the science of herbal medicine had become a drawing room curiosity, and homeopathy was for the cranks. Sadly the first part of the *British Herbal Pharmacopoeia* was not published until 1976, so a period of plant awareness had come to an end, for as the eighteenth century arrived, some of the greatest changes the countryside would have to endure were about to take place. It would never be the same again and the warnings had been given.

Flax

9
THE FLORAL HOLOCAUST

Every valley shall be filled and every mountain and hill shall be brought low, and the crooked shall be made straight and the rough ways shall be made smooth.
St Luke

St Luke in the *New Testament*, paraphrases Isaiah's words of the *Old Testament* several centuries earlier. Did either of these prophets know what would happen on the earth centuries later? Was it chance, an allegory for spiritual change, or were they prophets in the true sense of the word, for this is exactly what happened in eighteenth and nineteenth century England? How and where does this gloomy prediction of landscape change fit into our floral history?

By now in this multi-coloured strand of our floral history, there were the native flora, early denizens, and colonists together, as well as numerous adventives, each with a story to tell. Plants had been raised for food in fields and gardens, grown for pleasure and medicine, then listed in herbals that had been written on their properties and uses. The fourth phase of plant change illustrated on page four was about to begin. The next two centuries were to add new threads and remove old ones from this multi-coloured strand and Isaiah's curious prophecies were to be put to the test.

The English landscape at the beginning of the eighteenth century probably looked much as it does today in the north and west of England, a patchwork of hedgerows and stone walls, acting as parish or land ownership boundaries, ditches and winding roads linking farms and villages, and with bare uplands looking much as they did when Bronze Age man had finished with them. The country between south Yorkshire and south Derbyshire in the north, Buckinghamshire in the south, and Hampshire in the west was very different. Here was a landscape of open fields surrounded by common land heaths, comprising gorse and woodlands. They were sanctuaries for our increasingly beleaguered plants of woodland and heath.

At the turn of the century Ray, who died in 1704, had listed nine hundred species of plant. Were there any unlisted new arrivals? The fact is that there is no record, for there was little interest in our flora at this time. The Linnean system of classification had not taken the world by storm, nor was there any great incentive either. Linnaeus was a man who combined his love of order in living things with an appreciation of their beauty. He had been enthralled by the splash of bright colour created by gorse on Putney Heath when he visited it in 1736. In 1753 his book *Species Plantarum* reduced the plant names to two only, now known as the binomial system. This meant that, for example, the bulbous buttercup became *Ranunculus bulbosus* instead of *Ranunculus foliis ovalis serratis scaponudo unifloro*!

Writings of the time were very self-centered. A good example is Defoe's description of the landscape between Rochdale and Halifax. He was in awe, if not a little frightened by it. He does not record that valley areas were being used for water power and the appearance of the land was changing. In places like Halifax and Stroud the countryside was rapidly being eaten up, due to the need for water and access to coal near the sheep-rearing upland. Similar events were recorded in the Black Country, where the iron and steel industries were growing. Farmland round the navigable rivers was disappearing rapidly like the farmland in sheep rearing areas. Unfortunately these were the important things to write about. Observations on natural subjects were non-existent. There was a darkness over the land in more ways than one.

One tiny light in the gloom was the 'herborising' around London by Sir Hans Sloane, the owner of the new Society of Apothecaries Garden at Chelsea, now known as the Chelsea Physic Garden. He listed the plants which he saw but there was nothing new. Not long afterwards publicity for herbal plants came from a very unexpected source; John Wesley, the founder of Methodism.

This well-known preacher crusaded strongly against the evils of the artificial medicines being manufactured in the newly developed laboratories. His pamphlet *Primitive Physic* was aimed at curing most diseases the natural way, which he believed was God's way. Amazingly this man's thunderous oratory created only a minor stir. A doctor, William Withering agreed with him, for he had discovered, from an elderly lady in Shropshire the diuretic effect of foxglove leaves, which furnished a natural cure for dropsy. At least the foxglove will never become extinct.

Meanwhile another change had been slowly attacking the flora, already under siege from the rapidly increasing industrial developments. Farmers had started to fence in the fields to keep animals away from the food crops while they were growing. It was also the method of keeping them contained in one place for grazing. New fodder crops were increasing the number of animals which could survive the winter, which also increased the need for fencing in the animals. All this helped to increase food production for a rapidly growing population. It also brought about the Enclosure system, which eventually changed the nature of the flora and the pattern of our landscape for ever.

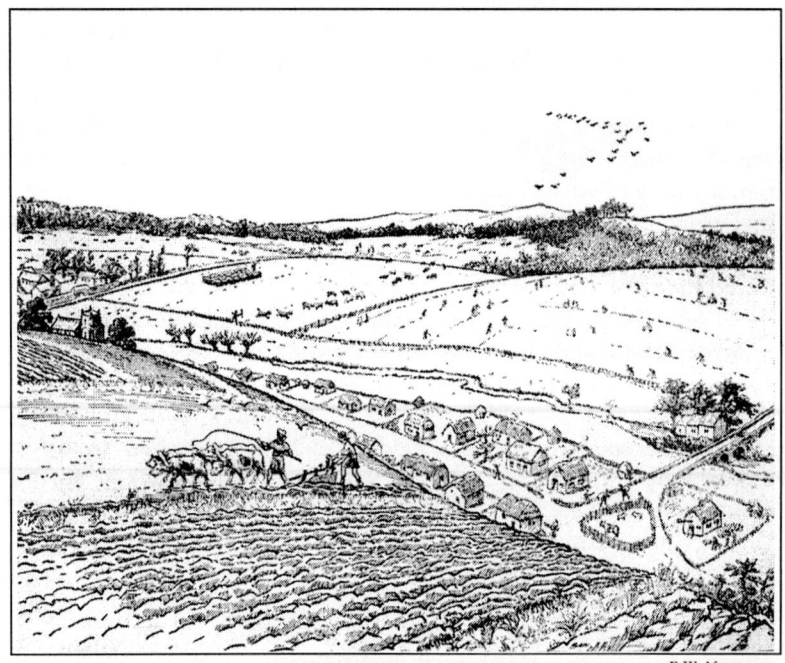

A pre-enclosure landscape

E.W. Neve 1922

Each villager had a greater or lesser number of scattered strips, which were cultivated in an annual rotation of wheat, barley, and fallow. The common land beyond the three large arable fields, known as the 'waste', was available for the communal grazing of geese, pigs, sheep, and cattle, as well as the collection of firewood. The agricultural methods suited a wide variety of fauna and flora in the headlands of each 'furlong' and the scrubland of the waste. The cultivated fields were sometimes protected from the free-ranging animals by movable wattle hurdles rather than hedges.

By the eighteenth century, if the owners of two-thirds of the total number of strips were in agreement, an Act of Parliament could be obtained to enclose not only the arable fields but also the entire commons. The land to be enclosed was subsequently parcelled out, according to the prior ownership of strips, and given over to sheep-grazing, with dire effects on plant life.

A post−enclosure landscape

The poorer villagers inevitably had to sell their allocation to pay for their compulsory share of the legal and fencing costs and left to find work in the towns. Their hovels were often swept away, 'to improve the view' from the manor house of the newly enclosed parkland, John Clare described the effect of enclosures on the landscape in some of the most descriptive poetry in the English language.

Inclosure, thou'rt a curse upon the land, And tasteless was the wretch who thy existance plann'd. Inclosure like a Bonaparte let not a thing remain. It leveled every bush and tree and leveled every hill.

In the early stages of the enclosures, the land which could not be used for grazing purposes was planted with trees. For a while John Evelyn's dream of a renewable resource was being realised, for the native trees were being grown for planting the new enclosure hedgerows. The vast areas of heathland, used since medieval times for grazing and cutting firewood by the rural population, were sadly lost for ever. So were the open fields and winding cart tracks ridden with such dedication by the early botanists Turner, Gerard, Johnson and Ray. The strip farms, now seen as ridge and furrow, were the last relics of John Clare's rural England, for the heathlands had already gone before he died. The land became the familiar chequerboard of fields, hedges and service roads which we know today. The last phase of plant change was under way as we moved from a total rural economy to an age of industrialisation and urban development.

History often repeats itself. In the eighteenth and nineteenth centuries there was the creation of a stony and barren landscape, similar to the one left when the glaciers began their last retreat at the end of the Pleistocene Ice Age. The combined effect of the Industrial Revolution and the enclosures was to produce the same thing, for quarries and gravel pits were dug, and there was exposed soil along the newly planted hedgerows. Some of our native plants could go 'back to basics', finding refuges in these newly created places.

There were some new plants joining the local inhabitants on these stony places. Mind your own business—from the Mediterranean—seems to have arrived in our gardens at this time, though some plant historians feel that it had arrived much earlier. It lives on stone walls today, forming a dense mat of tiny bright green leaves, but when established it outstays its welcome and can be removed by the bucketful. This is probably how it made its way out of the garden via the rubbish tip and onto an urban wall environment, apparently immune to pollution then and now.

Another mid-century arrival was the North American bridewort, a shrub with white or pink flowers. This soon escaped out of the garden. A central European member of the family also arrived and joined its North American cousin in the more attractive urban developments. Another arrival in the nineteenth century finally hybridised with one or both of these plants. The new plant being named, unsurprisingly, confused bridewort, can be seen on wasteland today.

The Floral Holocaust

Although there was wholesale destruction of habitats in the eighteenth and nineteenth centuries in the name of increased agriculture and industrial development, there were some positive events. One of these was the development of hedgerows. From 1750 the enclosures became a very rapid parliamentary process, which revolutionised the appearance of the countryside. The open vistas of Shakespeare and Clare were gone for good, but in their place were patterns of hedgerows. There were probably two hundred thousand miles of hedgerows planted between 1750 and 1850, about the same length as all of those planted in the previous five hundred years. Added to this were the miles of stone walls, built where the trees were unlikely to grow. These two features created two new habitat refuges for our flora, since plants will live on, in and around walls as well as becoming part of a hedgerow community.

Enclosure hedgerows had to be in place within a maximum of twelve months after the award had been made. Pre-parliamentary and early parliamentary hedgerows were planted with local trees, such as oak, ash, elm, blackthorn and all with hawthorn. Later plantings just used hawthorn which grew very quickly; it was not known as 'quicksett' for nothing. Initially a spinney—a small wood or thicket—would have been the source of hawthorn, but as demand grew, many millions of hedging plants were grown in the Midland plant nurseries, the area of highest demand creating fortunes for those who grew them.

In the east Midlands, the land of the Quorn Fox Hunt, the land clearance and enclosures took on a different look which can still be seen today. The spinneys and coppices were not cleared but left in place to give the foxes cover to raise their young. Here is the irony of encouraging a pest so that it can be 'sportingly' hunted. Another refuge emerged out of man's desire to hunt foxes. The hedgerows were adapted to the hunt as well. They were planted with local ash and elm, placed some distance apart with stretches of hawthorn between them. This was to make them easier and safer to jump because the hawthorn was lower.

As if the enclosures were not sufficient upheaval for the flora, the roads were altered as well. The curving bends of the cart tracks were progressively replaced by straight lines with the odd right angle around new enclosure fields. The map overleaf of the rectangular fields near Holbeach shows this very clearly. The small fields would have been cut from open pasture near the original ancient homesteads, now gone without trace. Interestingly, these often represent old furlong boundaries and trackways between one village and the next, which may be a nuisance when driving in a hurry.

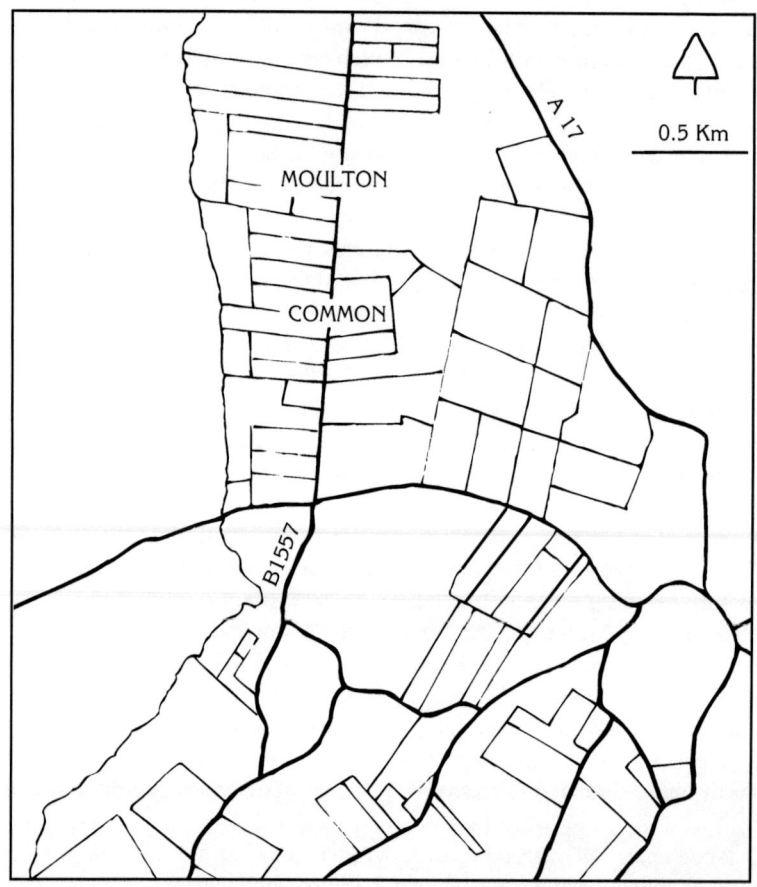

Rectangular fields near Holbeach

The width of the verges is quite large, since when the roads were marked out there was an allowance for moving sideways off the main track in bad weather. These wide verges are excellent refuges for the graceful white cow parsley, orchids, vetches, clovers and trefoils which thrive on the less fertile soil. Many of our native grasses have taken refuge here, adding their tall graceful shapes to the wide verges.

The Floral Holocaust

Today's maps of the enclosure counties still show some of these roads, which with modern 'no spraying' and controlled mowing regimes have become twentieth century havens for many of our rarer plants. Before the Enclosure Acts the main roads were less well defined, allowing the coaches to avoid the trouble spots, especially in winter and early spring. Just imagine the effects on the chalk flora of Salisbury Plain when it is appreciated that the London to Exeter road, which crosses it, was a quarter of a mile wide in places, so that the sticky chalky patches could be avoided. Verges suddenly seemed to be very respectable places. By contrast, road maps also show that some areas were unsuitable for enclosures. The maps of the roads around Boston, Lincolnshire, show that the roads meander around, connecting all the settlements and probably using the ancient pathways.

The country became financially richer and the early managed woodland was abandoned as it became cheaper and quicker to import wood for poles and pit props. John Evelyn's oaks matured to supply the dockyards with timber for Nelson's navy during the war against the French at the end of the eighteenth century. Would the sober and earnest, tree-loving John have approved of that twist of fate? At least we did win that war in the end. Unfortunately, there was no other forestry policy until 1916, when the Forestry Commission was set up.

However all was not lost, for more money meant more leisure time and more leisure pursuits, like hunting game—not foxes—and gardening. The eighteenth century passion for pheasant shooting certainly kept our small woodlands intact, for they were conducted on private land. They were well managed to give the pheasants, who were probably medieval introductions along with fallow deer and rabbits, good cover under which to lay eggs and raise their young. The flora survived well under these conditions and would have continued to do so if the rhododendron had not been introduced.

Around 1760 the rhododendron first appeared in the gardens, planted for its beautiful flowers and evergreen leaves giving colour in winter. This native of Turkey and Armenia was then used for pheasant coverts as well, which was when the trouble started. It proved impossible to control and took over whole woodlands, stunting the growth of other saplings growing in its shade. In Snowdonia it has proved almost impossible to eradicate, and has taken over large areas of land. It really is a plant thug!

The increase in available money allowed wealthy landowners to buy and enclose land. Whole villages including churches and productive cornfields were incorporated into huge deer parks. This was the era of the landscape gardeners like Bridgeman, Brown, Repton, and Kent, who swept away the natural vegetation and permanently altered the landscape—Isaiah?—to fit their designs. These flowerless landscapes pushed the wild flowers to the limit and there must have been many species of plants which could not survive in the new hedgerows and verges.

Capability Brown is credited with cutting down large numbers of elms, which further reduced the natural population. However, good use was made of local trees such as oaks, beeches, limes and Scots pine. New trees such as the giant cedars were introduced to give the parks a touch of individuality. Curiously, although they survived in their new habitat they have never strayed out of the park, and nor did other introductions such as the numerous maples from Pennsylvania. At least the native trees did not have to compete with these trees in the wild.

In spite of all the upheavals a couple of newcomers were recorded. At the end of the eighteenth century *galinsoga* arrived as a curiosity at Kew Gardens. Just as the Oxford Botanic Gardens proved unable to contain Oxford ragwort, Kew Gardens were simply not big enough and by 1860, Kew weed or gallant soldier—since its correct name *galinsoga* had been found to be unpronounceable—was recorded outside in the London area. It is not a sturdy upright soldier-like plant, being more like a wimp with a weak curved stem, small dingy white flowers and yellowish green leaves. Curiously it often occurs in the market gardens close to London, which used grey shoddy waste from woollen mills for manure. Since wool had been imported from South America from the early nineteenth century, gallant soldier may have had a less glamorous route here on the unwashed fleeces of South American wool. It has also been recorded on Lake Garda and in Costa Rica; obviously a plant with an urge to travel.

It is worth remembering that at the end of the eighteenth century market gardens covered an area of some ten thousand acres in Middlesex alone. These hand-dug plots provided spinach, radishes, cabbages, cauliflowers—brought here by the returning crusaders—and other salad vegetables for the London market. This created an enormous number of rubbish tips. In addition there was little or no sewage disposal, so that the streets had middens of vegetable waste as well as heaps of household rubbish. Curtis in his *Flora Londonensis* lists many ruderals, all of which were listed on the rubbish tips of Mesolithic man, as being present on these eighteenth century waste heaps in the city. We have certainly provided our flora with a range of habitats.

The second recorded arrival is a less well known plant, pirri-pirri bur; a relation of salad burnet from Australia and New Zealand. Its arrival was probably by the same means as gallant soldier, on the fleeces of imported wool. There is no valid record, but it is often suggested that it was a garden escape. It may have been a simple hitch-hiker, but it probably reflects the country's trade in wool. Today it is found on sand dunes, where it withstands the trampling of holiday feet and still manages to spread.

During the first half of the eighteenth century there was no handbook to make flower identification easy or accurate. It would have been invaluable in an era of increasing money and leisure time, for many plants could have been recorded. Perhaps man cannot progress on all fronts together and must do one at a time. It is possible that life in the eighteenth century was too involved with industry in its many forms to care. Towards the end of the century there were more books, and 'herborising' had become popular, especially with women. By the turn of the century there was Sowerby's *Botany* in thirty-six volumes with a masterly text in English by James Smith, the first President of the Linnaean Society. At last plant worship on the lines of Ray's 'rich array of springtime meadows' began to be a respectable attitude. The next step was to see what had survived the holocaust.

Greater Plantain

10
FOR BETTER, FOR WORSE

I wandered lonely as a cloud that floats on high o'er vales and hills,
When all at once I saw a crowd, a host of golden daffodils.
William Wordsworth

The beginning of the nineteenth century saw the Romantic Revival in music, art and literature, with authors like William Wordsworth describing the countryside. Appreciation of the plant world increased together with a need to preserve it. To that end, in 1810 Wordsworth proposed that the Lake District should become a National Park. Suddenly there was an awareness of wilderness beauty, just as it was being indelibly marked by dams and weirs, roads and hedgerows. John Clare had forecast the future correctly when he described with some emotion, the permanent effects that the enclosures would have on the landscape.

The slowly declining canals, whose great cuts had scarred the landscape since the late eighteenth century, began to create large new habitats along the walls and paths of their reservoirs. This was an urban flora which could survive the pollution and trampling it was receiving. Native plants such as the greater plantain—the Anglo-Saxon waybread—thistles and dandelions had found new homes. The less robust water species also found new homes along the canal compensation reservoirs.

These were plants such as the flowering rush, bur marigold, and the great reed-mace, not bulrush, in spite of the Victorian artist who painted Moses in a bed of this plant! There was codlins and cream among the common reed which soon arrived, joined by plants around the margins such as water forget-me-not and lesser spearwort. Did the eighteenth century denizen, mind-your-own-business, survive so successfully because it found a new home on the walls of the canals, along with seventeenth century yellow corydalis

This calcicole came from the rock gardens of the southern Mediterranean region. With its drooping tassels of bright yellow flowers and olive green leaves, it escaped over the garden wall and happily joined other plants on the increasing number of brick walls. Today it grows readily with soldier's pride or red valerian—another sixteenth century arrival—and the ubiquitous ivy-leaved toadflax.

Part of the Agricultural Revolution had involved the import of uncleaned fodder crop seeds. Imported clover probably brought with it common or bird's eye speedwell. A native of south west Asia, it is now found around the edges of cultivated land from fields to gardens, adding a splash of blue colour. This is not always appreciated when it has to be dug out of the flower bed, for it is a very persistent weed—or a wild flower in the wrong place?

Gardeners have always lived in a world apart, relatively untouched by social upheaval. A war in Europe did not stop the Mediterranean winter heliotrope from arriving in our gardens. It is a relation of coltsfoot and as hard to eradicate from the flower bed. Its native cousin butterbur had an original use before refrigeration was developed. Country people wrapped their butter in the large umbrella-shaped leaves to keep it cool. History does not relate if the leaves of winter heliotrope were used in the same way, but it joined cousin butterbur on verges and by streams where it flowers today.

Thanet or pepper cress arrived a short time later. Britain had sent soldiers on an ill-fated expedition to Waldcheren Island in 1802. Injured soldiers were brought back home on hay-stuffed mattresses and landed at Ramsgate. The soiled mattresses were sold to a farmer on the Isle of Thanet, who dug them in as manure like wool shoddy. Enter Thanet or pepper cress whose seeds had been in the hay of the mattresses. A plant of central and southern Europe, it is now widely distributed along the roadsides and on waste grounds in England. It was recorded in New Zealand in 1895, arrived in Russia during World War I with animal fodder and has since been recorded in the United States and in Australia. How did it move around the world? There are recordings of the pepper cress or Thanet cress on the banks of the river Tawe near Swansea. This could have come in on ballast stones, so there may be at least two methods of transport and there may be at least two species.

Another wartime arrival was the monkey flower. It is a native of the gloomy Aleutian Islands, where it rains on average for two hundred and fifty days each year. It must have felt at home here—is our climate so wet?—for it had made it out into the wild by 1820, growing by rivers, streams, ponds and canals. A relative who arrived a little later from Chile, blood drop emlets—because of the red spots on the petals—also made a highly successful and rapid bid for freedom. The slowly declining canals of the eighteenth century were brightened up by the mass of bright yellow and orange nasturtium-shaped flowers colonizing their dreary banks and drains. They both will colonize in rural settings. The lovely Upper Teesdale valley below High Force has monkey flower growing with the betony and Teesdale violets.

Napoleonic wars or no Napoleonic Wars, enthusiasm for new garden plants was undiminished. Snowberry with its round white berries and olive green foliage arrived from North America in 1817. A wild fuchsia *F. magellanica* reached here around 1820 from South America. This was not the first fuchsia, for Kew Gardens had a such a plant in 1788, though a different species *F. coccinea*. The South American arrival soon escaped from the garden to far more homely surroundings on the marshy land outside. Orange balsam arrived at this time, how is unclear. It is native in the Maritime Provinces of Canada and New England, and had established itself a permanent place in our flora by 1822. The Americans call this plant 'jewelweed', as well as 'celandine', since it was used to cure warts. The Pilgrim Fathers had taken the greater celandine, a Roman introduction into our flora, as a cure for eye complaints and warts. It makes its name a little confusing so let us keep to 'jewelweed'.

Fuchsia and snowberry were joined in the wild by another less attractive garden escape called Japanese knotweed. In 1825 it was imported both as a fodder crop and as a garden perennial. As a fodder crop it seems to have been a non-starter, but as a colonizer it was extremely good. An amusing Cornish story of the 1930s tells of 'Hancock's Curse', which was a garden so full of 'Jap weed' that it spread around the area including other gardens. This caused one house to lose £100 from its sale value; the price loss due to the garden being smothered by this weed. That was quite a large sum of money at that time.

The woollen mills of Yorkshire imported fleeces from North America as the demand for woollen cloth increased. In the discarded shoddy were the seeds of the spiny cocklebur, which joined the band of early nineteenth century arrivals which had found homes in our new and increasingly industrial landscape. It made its home with snowberry and fuchsia in hedgerows and damp places round the edges of canals and waterways. It also joined the new colonizers Japanese knotweed and orange balsam on the numerous areas of waste land.

For Better, for Worse

1815 was a watershed for European life including ours here in England. Once Napoleon Bonaparte had departed for St Helena we came alive again. Travel became easier and the quest for new experiences suddenly took off. It was in this period of discovery that the next milestone appeared. In 1825 the first passenger railway between Stockton-on-Tees and Darlington was opened. The invention of the steam engine and the development of the railway network added to peoples' ability to travel both at home and abroad. By the second quarter of the century the early botanists plodding along the winding pre-enclosure tracks on horseback, had been replaced by top-hatted Victorian gentlemen with their long-frocked ladies, who arrived by train. They travelled to see the floral wonders of places as far apart as Cornwall and Teesdale, clutching one of the easy-to-use flora, in English, by such authors as Lindley, Hooker or Babington.

At long last, field botanists were able to choose from a selection of reliable and up-to-date handbooks, with which they could enjoy their flora. After fifty years the nomenclature of the Linnaean system, married to the natural classification of Ray, had produced an arrangement which, with only a few minor modifications, stands today. These major floras were not the only ones, for the amateurs, in true camp-follower style, had produced a range of local floras ranging from species lists to accounts of the local climate and geology. They were in English of course, but with styles which varied from brusque scientific prose to ludicrous expressions of sentimentality. An example of the latter:

> *Whether presenting a bouquet of flowers with courteous style to his lady love, or moralising as he hangs over the topmost turret of a princely ruin to pluck a sweet gem . . . the Botanical Looker—out is ever at home.*

This would have been no use at all with an unknown plant especially if it was raining!

> *I am a tall annual plant, with purple—pink to white flowers, which appear from July to October, and I grow round the banks of streams and rivers.*

This description of Himalayan or Indian balsam, or policeman's helmet was written by a little girl after a visit to the local nature reserve. It was introduced as a hot house plant in 1839, becoming a troublesome weed of waterways before the turn of the century. The flower shape, size—it is often a metre in height—and unique, explosive seed dispersal mechanism had made it a fascinating water plant to the small observer on a 'Watch' Saturday morning outing. The reason for its highly successful escape from the greenhouse and the garden is due to this explosive seed dispersal mechanism. This allowed the seeds to be carried along numerous river systems and artificial waterways as well as on the air currents, like its cousin orange balsam. This plant as its name suggests comes from the Himalayas. There it grows on the banks of streams fed by the glaciers, so our relatively warm waters must seem very good to this Himalayan plant. That area of India was once part of the British Empire which begs the question, would it have been brought here if India had always been independent?

By 1852 four newcomers had been recorded in our flora, the dubious benefit of increased travel. Canadian Pondweed or waterweed arrived here via County Down in Ireland in 1836. Claridge Druce, an eminent botanist of the time, recorded its introduction into the Oxford Botanic Garden and the event which followed. The name 'drain devil' was a well deserved nickname, for a plant which blocked the drains and streams so successfully in the Oxford area and beyond. Spring beauty arrived around 1852 from the Pacific coast of North America as a substitute for spinach. Escallonia, an earlier arrival from the South American island of Chiloe joined spring beauty over the garden wall. The sticky leaved escallonia was brought here as a hedging plant, but today its pink-red flowers beam out from many areas of unimproved land near the sea in south west England. The fourth arrival was our third and last balsam from the Russian regions of Siberia and Turkestan. Perhaps its somewhat shabby appearance and insipid yellow flower is related to its harsh homeland, for it lacks the luxuriant charm of jewelweed and policeman's helmet, which is reflected in its name of small balsam.

By 1851, the year of the Great Exhibition, there was a well-developed railway system created at enormous cost to the woodlands. Until the nineteenth century, large woodlands had not changed a great deal, for it was the heaths and copses on common lands which had suffered from the enclosures. The railways cut huge swathes through the countryside, felling an enormous number of trees without a thought for tomorrow. The rest of the damage was done by the 'navvies' who laid the lines, dug the tunnels and built the viaducts. Think of the service areas laid waste when there is a motorway extension, using the supposedly more efficient modern machinery. The Victorians had to add to this equipment area, the living accommodation for the navvies! This created such havoc in some places that Wordsworth entered the argument against railway development. He had become a formidable conservationist, especially when his beloved Lake District was involved. He was a vociferous objector to the proposed Lake District Railway Line. It is a nice thought that the author of 'Daffodils' may have started the environmental protest movement.

It has happened before, for out of a disaster came something good. This time it was the new railway embankments and cuttings, fertilised by the ash from the coal-fired steam engines. A succession of pioneer plants gradually covered these new habitats, proving once again that our flora cannot be subdued. It just finds a refuge, then comes marching back out when the time is right. What is more, the plants recovered the trampled and mangled earth left by the construction gangs. These were the plants which had been here since the glaciers finally retreated, covering a land which was desolate heaps of stone and rubble, then as now. Plants included grasses, stinging nettles, knotgrass, chickweed and mugwort, as well as the other ruderals which are always colonizers of waste land goosefoot and orache. New arrivals like the clovers appeared to stabilise the soil leading to the arrival of vetches and plantains. The new meadows and finally copses of the nineteenth century were arriving.

The air currents created by the trains carried the light seeds of birch and alder, early colonizers as they had been in the Great Rock Garden. These were joined by herbaceous plants like primroses—those Neolithic arrivals and safe here from the Victorian gardener's trowels; dandelions, thistles, coltsfoot, native and Oxford ragwort, groundsel and ribwort plantain, all arriving here as the air currents carried their light, wind dispersed seeds. Newcomers to the flora such as ivy-leaved toadflax, golden rod and later on buddleia were spread around the country along with the natives and earlier newcomers.

The *compositae* or daisy family is well represented in any habitat where there is considerable wind dispersal of seeds. Many have found permanent and safe refuge on the banks around railway lines, sidings and marshalling yards. Others in this group include yarrow, Kew weed or gallant soldier, Canadian fleabane, tansy and mugwort. Perhaps a favourite of many people is coltsfoot, not for its rather undistinguished yellow flowers but for its roots. They were often soaked and boiled in sugar to make a type of rock. Then they were given to children to suck, relieving the discomfort of sore throats and often tonsillitis. It was a useful plant all round, for it made excellent cough medicine, beer, jelly and wine. The hairs on the large leaves were scraped off and soaked in saltpetre, for use in a tinder box before matches were invented. Boiling the plant to make a poultice finishes off an impressive list of uses, yet it still flourishes as a plant on wasteland; a true plant survivor.

Sadly, there was considerable interference in the natural cycles of our woodlands. They were treated first to care and then to neglect, boom and then bust. Oak bark had always been used for tanning leather, as well as the wood being the chief building material for ships. From 1780 onwards the demand for leather and ships rose sharply; it was that man Napoleon again. After Waterloo in 1815, the demand slowly declined and by 1850 it had collapsed with development of iron ships, leaving a comparatively small demand by the leather industry.

This completely disrupted the management of the timber woodlands, leaving the standard trees to take over. Few understorey trees were to survive in the decreased light. The natural canopy and shrub layer state had been lost. The flowers suffered too, for the plants of the woodland floor now had little or no light when the trees were in full leaf. The agricultural boom of the 1840s had removed more ancient woodland to add to the losses created by the development of the railways. The heaths had not been grazed at this time and they regenerated into secondary woodland. By 1880 the agricultural slump created yet more secondary woodland as the cultivated land was abandoned, but those ancient woodlands removed in such a hurry could never return. The land gained more new woodland but lost the old, something that could never be replaced.

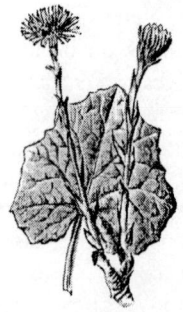

Coltsfoot

9. *Wild daffodils near Bromsberrow*

10. Feverfew

11. Yellow Flag Iris

12. *Wild Arum*

11
THE CARPET BAGGERS

We are not rootless vagabonds, we are on our way to power.
Paddy Ashdown

The beginning of the twentieth century is a useful point at which to look backwards, to see which plants have become established in other nations' flora, for it must be obvious that plants are natural wanderers. Many will not stay in one place; plant them in your garden and they soon escape out into the wild. They will roam freely like true vagabonds, becoming denizens in a new land, like ivy-leaved toadflax and policeman's helmet. As well as being vagabonds, plants are natural hitch-hikers. Give them transport and they will travel the world with us. They have travelled with animal fodder, on clothes such as Roman sandals, on ballast stones, in crop seed and in the stuffing of mattresses and saddles, to name just a few of the known plant transport systems.

Exploration, trade, and war were the original starting points for vagabonds and hitch-hikers alike, though tourism could now be included in the twentieth century. In 1995 ivy-leaved toadflax, first recorded in England in 1618, was found on a wall of a Hindu temple on the island of Bali. It was a very long way from its native home on the rocky Mediterranean shore. This shows that there is another thread in the multi-coloured strand of our floral history; those plants which we have handed on through exploration, trade and war, and which became colonists, denizens and adventives around the world.

Periods of social and economic stability provide an opportunity for new areas of learning. This was the situation in the Tudor and Stuart periods of our history. The time of calm led to the planning and carrying out of expeditions to far off places such as the Caribbean, as well as North and South America. This new aspect of life was recorded by generations of newly literate people. By 1621, the Pilgrim Fathers had established a thriving colony in Plymouth, Massachusetts. This started a new wave of settlers leaving England and parts of western Europe for the New World, the eastern seabord of North America.

Enter John Josselyn, Gent., the seventeenth century author of *New England Rarities Discovered*, published in 1672. Little is known about the man's background. All that is known is that he visited his brother Henry in Maine, New England, in 1638 and 1663. This book is an account of his observations in the area around Black Point and Scarborough. Among the things which he recorded, in an accurate and sometimes amusing manner, were English flowers whose medical uses he obviously knew very well. He was apparently a man of wide learning, a pity that so little is known about him.

He lists three plants, dandelion, stinging nettle, and greater plantain, which would have probably arrived by accident, since they have no obvious uses. Their means of transport would have been in the bowels of the animals, fed on hay containing the seeds of these plants during their voyage across the Atlantic. North American plant studies have shown that the dandelion is not a native species, so this must have been its first recording as an adventive. Josselyn also records that the stinging nettle was the first English plant which he saw, and the large flat leaves of our greater plantain earned it the native Indian nickname of 'Englishman's or white man's foot'.

The settlers, records Josselyn, had taken plants, probably as seed, for household and medicinal use. This was exactly what the Romans had done when they invaded England. Dyer's greenweed had been taken to provide a yellow dye and, like a true vagabond, it escaped over the garden wall. It now covers the stony hillsides of the State of Massachusetts, completely naturalized in the New World flora. Ground ivy, known then as 'alehoof', was taken to be used as a substitute for hops and for cleaning. White and purple comfrey or comferie, often called 'knitbone', was grown for poultices to put on sprains or broken bones,

Tansy a medieval arrival here in England and also on John Josselyn's list, seems to have been grown for sentimental reasons, as well as its use as a colouring agent in cakes, as a moth repellent and to prevent miscarriages. Stinking mayweed may have reached New England by mistake, since it is a crop weed. This is a plant which would have been better left behind, since it cuts the hands like tares at harvest time. Josselyn lists it as ironwort, which is a good name for a plant with a tough stem. He also mentions that it was grown to make an unguent or ointment for sores. After several hundred years in the English flora, having been brought here by the Romans, houseleeks, and pennyroyal went on another journey to New England. Josselyn records the houseleeks as a protection against thunder and pennyroyal as an excellent cure for coughs including whooping cough, the gripes, stones —where?—jaundice and childbirth. How is the last condition 'cured'!

From other contemporary literature, it is recorded that many more plants became established in the New World flora either as colonists, for example stinking mayweed; denizens, for example dyer's greenweed; or as adventives such as greater plantain. It was the gardens and the vagabond trait of the plants which together produced the large number of New World denizens. The intriguing point is that the majority were not native English species. Are the vagabonds of the world a more hardy type of plant?

The Roman greater celandine and the medieval European soapwort are good examples of denizens and vagabonds both in England and in the New World. The latter came to be used as an antidote to the local poison ivy as well as a washing aid. One of our native poisonous plants green hellebore, was taken to the New World and used with great care in the treatment of cattle. It is a rare denizen with no vagabond tendencies this time. Non-poisonous English natives such as viper's bugloss and marjoram were used for a variety of ailments from depression and lumbago, to a general cure-all. Were these cure-alls just placebos for people who were genuinely homesick? It was a huge change for these settlers in New England, as it had been for the Romans in England centuries earlier. It was certainly why the thrifty Scottish settlers planted heather, for its look of home as well as a mattress filler. This is a denizen with a huge vagabond capability.

Pure pleasure was the reason for planting musk storksbill with its dark flowers and sweet smell. Another attractive plant 'butter and eggs' better known as toadflax, was planted in New England gardens for its bright yellow and orange spears of flowers. Star of Bethlehem, possibly a non-native here, was yet another plant taken for the pleasure of its flowers. These three plants have spread rapidly outside the garden in the countryside of the New World, blooming as vagabonds and denizens in the Land of the Free. Finally a less well known crop plant, buckwheat, was taken to the colonies to be used on the poor stony soil of the northern states. Here in England it was a poor substitute for wheat, grown on the less fertile soils of northern England. This has left the fields, becoming another plant from England to establish itself in the North American flora.

A botanical curiosity of considerable legal importance arrived in the New World at this time. This was hemp, which had been grown in England since medieval times. The settlers in Maryland and Virginia were given hemp seeds to grow so that the resultant rope could be traded for goods from an increasingly industrial England. Unfortunately this did not work out, for as Burns put it 'The best laid schemes o' mice an' men gang aft a-gley'. The settlers only grew sufficient for their own ropes and grew the much easier-to-raise crop tobacco. The profit margin was considerably greater. The hallucinogenic property was known on either side of the Atlantic, but not used to any extent. As a true vagabond it made a very successful move in to the wild, so by the twentieth century it had covered nearly 150,000 acres. There had also been a drug exchange of considerable significance, marijuana for nicotine.

It has been estimated that forty percent of the 'weeds' common in the State of New York are British. Apart from the plants mentioned with the New England settlers, there are Canadian fleabane and golden rod —did we just send them back? The 'vegetable rat' charlock with its bright yellow flowers; English natives such as shepherd's purse, curled dock and creeping thistle; the Roman sowthistle, mugwort—for clothes presses?—yarrow, red clover and coltsfoot—for sore throats and tinder boxes, and the Neolithic fat hen and Thanet cress. The last-named only arrived in England in 1809 so this island must be a rapid transit camp for some plants, which begs the question; what is en route now?

While the flow of emigrants to the New World continued, explorations were taking place further west in the Pacific Ocean. Captain Cook had rounded Cape Horn, found Tahiti and New Zealand finally landing in Australia in 1770. In spite of a less-than-friendly welcome from the aborigines in Botany Bay, Arthur Phillips set up a penal colony in nearby Sydney Cove in 1788. This started a wave of settlers to a rich and underdeveloped land. By 1829 Great Britain had claimed sovereignty over all that huge continent. Although by 1853 no more convicts were sent to Australia, there was still a high immigration rate, especially from Scotland. The Gold Rush of 1856 increased the flow of immigrants even further.

Along with this flood of people came the plant colonists, denizens and adventives to the newly discovered continent of Australia. In the crop seed sent to farmers there was knotgrass, known there as wireweed because of its tough stems, and black nightshade. Our native common ivy has become a denizen of considerable frequency and a nuisance in some places. Again as in New England, golden rod and blue periwinkle have escaped from the garden and become Australian denizens, joining fellow English plants such as the originally Roman hemlock and fennel denizens all. The list of adventives includes our Neolithic fat hen and charlock, along with the Roman scarlet pimpernel and sowthistle, the Dutch white clover and Canadian fleabane completing a short list of some of the plants which arrived probably in the crop seed or grass mixtures.

There was a lack of suitable grass for grazing or hay on which the British cattle and sheep could be raised. This had meant that the settlers had sent home for native species such as perennial rye grass and sweet vernal grass. Inevitably there were hitch-hikers in these British mixtures; waste land grasses such as meadow grass and fescue, the latter being a nuisance on the sheep's wool. Did quaking or totter grass arrive as a decorative garden plant or did it hitch a lift with a grass mixture? Which ever way they came, all these native British grasses are now well established in the Australian outback.

Gorse was a deliberate introduction to be used for stock control, since it was cheap and easy to grow. It grew so well that it soon escaped and today gives the landscape the same warm yellow glow that had so entranced Linnaeus on Putney Heath in the eighteenth century. The English broom and sweet briar grow with the gorse in the Australian countryside; were these denizens or a pair of energetic hitch-hikers? Is blackberry to be added to the denizen or adventive list for it is an excellent hitch-hiker? Plants which are definite adventives and did not make the journey to the New World are creeping buttercup, pellitory, ribwort plantain, petty spurge, St John's Wort, chickory, spear thistle, and common vetch. Did they get there in the packing cases, crop seed, or in the bowels of the animals taken out to Australia in the early days of settlement?

New Zealand was discovered by Captain Cook before Australia, in 1769. It was never used as a penal colony and was rapidly settled in spite of Maori objections. The hitch-hikers and garden vagabonds which grow there are very similar to the Australian ones. Thanet cress made it out to New Zealnd in 1895 and a more troublesome plant, ragwort, arrived around the same time. The native and poisonous English ragwort is a food plant for the caterpillar of the cinnabar moth, which controls its British distribution. Robert Burns preferred to remember ragwort's more traditional function, that of an economy class, fairy broomstick. Neither of these functions is of any consolation to the New Zealand farmer. It is a very common poisonous pest on their pastures, where it is neither controlled by caterpillars needing their lunch, nor by a population of fairies needing a ride. Today in England there are measures in place to eradicate it from our grazing land, where it has also become a serious pest. How it reached New Zealand is not clear; feeding stuffs or crop mixtures are the most probable means of transport.

South Africa had been on the early trade routes to the east since Vasco da Gama rounded the Cape of Good Hope at the end of the fifteenth century. Along with other European countries the British had settled the southern part of the continent. It is logical to assume that there would be traces of these settlements and cultural influences in the flora. However this is not so, probably due to the climate. New England, Australia and New Zealand all have climates not vastly different from Europe, providing the plants which arrived there with comparable conditions for growth. They could literally put down roots and become part of a new society of plants. South Africa's warmer and drier climate would not provide as many niches in which the European plants could survive. All that can be seen of north European origin is a collection of the permanent colonizers of the world's major cities which are curled dock, chickweed, stinging nettle, and the Roman sowthistle.

No look at hitch-hikers abroad can be complete, without a quick look at what has and is still arriving here in England. The urban flora is usually the first place to check, especially near dockland areas, for these are usually the first home for adventives. Another less attractive place is the rubbish tip. Various overseas foods can be seen growing here such as mung beans and date palms to name but two. Birds, especially migratory species, bring in seeds on their feet and discarded bird or chicken seed will also find its way there and start to grow. Sadly all these plants from warmer lands never survive in our north European climate. It is rather like South Africa in reverse.

Knotgrass

12

ABOUT THE HOUSE

When he has learned that bottiney means a knowledge of plants,
he goes and knows them.
Charles Dickens

Today the United Kingdom society is totally indulged, for it can buy almost everything which it needs in tins, jars, bottles, or frozen. If people are hungry they can go to the supermarket, there is the doctor and the National Health Service if they are ill. If clothes are needed, there are numerous shops. But it was not always like this, for each would-be-householder had to provide everything for himself. He had to find his own building materials before he could build his home, and make the furniture and utensils to go inside. Every housewife knew the plants which she had to collect for immediate use or for drying and storage, apart from those which she had grown for food. Country folk in rural continental Europe may still be seen collecting wild herbs and other natural foods today.

Before the floral story moves on to the twentieth century, another look at our plant companions reveals how the usage of some of them has resulted in their being well managed and protected. After all, the flower beds, gardens, and orchards were very safe refuges. Another thread in the multi-coloured strand of our plant history suddenly emerges. It draws together how man lived side by side with the plants which for him, had a definite use.

Plants about the house can be divided into several groups. The first and most obvious group contains the plants used in house building. Only those potential home owners with money could afford to build in stone and later on in brick. The majority of homes until the Industrial Revolution were built with mud and wood. The story of the oak has been told but hazel was also used, whose whippy branches were ideal for creating walls of wattle and daub around an oak frame. Hazel coppices are here today as part of our landscape history and in regular use, for hazel is still used in the twentieth century for sheep hurdles. An odd use for hazel was the production of hazel charcoal, which when combined with sulphur and saltpetre made gunpowder. It is assumed that this would only be a domestic activity in time of war.

Birch, one of the oldest Ice Age survivors, was and still is a common tree. It had many uses, as the following nineteenth century account by J.C. Loudon will show. 'The highlanders of Scotland make everything of it; they build their houses, make their beds and chairs, tables, dishes and spoons, construct their mills; make carts, ploughs, harrows, gates and fences, and even manufacture rope with it.' Other uses not recorded by Loudon included thatching, tanning leather, and at a pinch, as a mattress stuffing.

Another group of trees included those whose wood was used in the furnishings. A tree in plentiful supply was the sycamore, that curious adventive which grows so quickly around houses and gardens, even in the poorest of soil. The hard-wearing, creamy-coloured wood was used to make kitchen utensils and the beautifully carved Welsh love spoons. Native beech was carefully managed to provide a supply of kitchen chopping blocks which were regularly cleaned with sprigs of butcher's broom, cut from the hedge for a medieval scouring pad.

Some herbaceous plants were used in household cleaning and cooking. Horsetails were gathered for cleaning and scouring, in the same way as butcher's broom, since the silica spicules in the stems would clean up metal-ware and smooth off the sides of arrows. Lady's bed-straw joined birch as a mattress stuffing, as well as a substitute for rennet in the production of cheese. The starch in the roots of lords and ladies—wild arum—is very fine. This made it easy to use when thickening milk to produce a drink called saloop, a popular drink for working people first thing in the morning. It disappeared when tea and chocolate became common and cheap to buy. In Tudor times it was used extensively to stiffen the very fashionable, huge ruffs, collars and cuffs. It was often sold as Portland Sago, available in shops on the Isle of Portland as late as 1880. Unfortunately it was very damaging to the hands of the laundry women, causing cracks which were slow to heal. This problem had been recorded as far back as John Gerard's time, but in the absence of any Health and Safety legislation, the starch was used until the end of the nineteenth century. The plant seemed to grow in such profusion, especially on the Isle of Portland, that it was rarely cultivated.

Today a plant-strewn carpet or house floor would mean the housewife would get out her vacuum cleaner. In the days of earth or stone floors, there were plant carpets of rushes which were very ecologically friendly. They were renewable, biodegradable, and had strewing herbs added to them to hide the smell of uneaten food, animals, and general rubbish. Gerard always recommended the native meadowsweet along with sweet flag. This arrived here in England in 1567, a useful plant with sturdy stems and a sweet pineapple-like smell. By 1610 it was being regularly cultivated but inevitably it escaped out of the pond, over the garden wall to the streams and rivers beyond. It was so abundant in Norfolk that the whole of Norwich Cathedral floor was covered every summer with this aromatic rush. Some churches in the nineteenth century are recorded as delaying the laying of a stone floor because of the readily available, cheap alternative. Apparently the churches were short of money then as well as today.

If meadowsweet and sweet flag are added to a list containing Turner's melilot, common fleabane, and woodruff, there was a range of easily available sweet-smelling plants for use as strewing herbs. Records show that all these plants were cultivated to some extent with the exception of woodruff. It contains an aromatic compound coumarin, which is a basic perfume fixative and explains its use in clothes presses and mattresses, along with common fleabane which was probably always just an insect repellent. Tutsan was also used in bedding and clothes presses. Hops, then as now, were used in pillows to cure insomnia.

A look at the list in the appendix entitled 'Some common medicinal plants', shows that many of the plants such as stinging nettle, coltsfoot, ground ivy, and pot marigold had more than one use. These plants were often picked in the wild and transplanted into gardens. Then as now, there were favourite remedies, which explains why plants are less common in some areas than others. This regionalisation may be due to climatic and soil differences as well, but local tastes also play a part.

Betony, a common garden plant, was used to make a completely drug-free medieval tonic wine, the equivalent of present day Wincarnis and Sanatogen. Henbane extracts were a common sedative and pain-killer, being recorded as easier to use than hemlock, but how did the early herbalists know this fact? The safest and most commonly used diuretic was dandelion. Was this plant, the bane of gardeners today, cultivated or was it just picked in the wild? Since it was so safe to use, it was used frequently on the continent, where its French name 'Pis-en-lit' indicates its efficiency. Foxglove, another excellent diuretic, is forever linked with Dr William Withering who first used it to cure dropsy. After this it was regularly cultivated for its effect on heart beat as well as fluid retention. The diuretic cordial from broom—the *planta genista* of Henry II—was a favourite of Henry VIII 'against surfets and diseases thereof arising'. Considering the excesses of his reign, whole gardens of broom would have been needed to produce the gallons of cordial, which the king would have required. There are no records of this level of planting, so perhaps his Majesty found another diuretic?

Did the need for invalid food contribute to the decline in numbers and distribution of the early purple, green-winged and military orchids? A saloop-type of drink was produced from the roots of these plants, which always grow very slowly. The roots contain a soluble, nutritious gum called bassorin. When it was added to milk, cream, and eggs, it made a nutritious and easily digestible drink. Records do not show that any orchids were cultivated for food. The orchid culture for the pleasure of the flowers only, came later with the development of heated greenhouses. Today the wild orchids are protected by law and digging them up is an offence.

Another group of herbaceous plants was used to dye clothing. Some of the English dyeing techniques were learned from the Flemish weavers, who traded in the wool areas of East Anglia in the fourteenth and fifteenth centuries. It is probable that their skill of mordanting was acquired at this time. This increased the colour range by the addition of aluminium and potassium sulphate or alum.

> *Tramp up Snowdon with your woad*
> *Never mind if you be hailed or snowed on,*
> *Never want a button sewed on,*
> *Go it, ancient B's.*

This rhyme suggests that blue dye from woad was the earliest and most famous English colour. The mystery is how did the Ancient Britons extract this dye, for it is a difficult one to collect? Elizabeth I banned its production when she was passing through a dyeing area, since it was a very smelly process as well as a tedious one. Once common in the wild and a valuable cash crop, woad is now very limited in its distribution. Fortunately, like hemlock, today it has found safety on the banks of the M5 Motorway. It had the privilege of being the first dye for policemens' uniforms, a combative colour to the end. It was eventually replaced by imported indigo, which was cheaper and easier to use. Another body colouring was extracted from gipsywort. The black dye was used to disguise people as Arabs or Egyptians, when posing as fortune tellers at fairs, or for those who were evading the law.

Good green dyes were only produced when blue dye from woad was added to yellow dye from weld or dyer's greenweed. Kendal and Lincoln green were regional variations, probably due to different mixes of the dye plants, possibly because one was more abundant than the other. Weld seed imported from Europe, was planted and grown as a cash crop along with imported madder seed, whose roots produced a red dye. Both plants inevitably escaped from the fields and became naturalized as denizens over a wide area of the country. Native dyer's greenweed is abundant especially on limestone. It used to be collected by hand in the wild and sold by weight. It was taken to the New World by the Pilgrim Fathers where it escaped from the gardens to become a denizen in a foreign land.

Finally there are the lichens which can be regarded as the most specialized of the dye plants. Their considerable use in Scotland, the north of England, and in Wales is due to their abundance in the moist climate, where they have few plant competitors. They produced gentle browns and greys and brighter colours, such as purples and reds, when mordanted with ammonia.

Our floral history is now a well developed, multi-coloured strand. A sideways look at our plant companions, both as fellow world travellers and about our home shows how close man has been to the plant world. The huge changes which started with the eighteenth century enclosures were to increase rapidly into the twentieth century. Their effects were to be even more profound than those created by the Romans.

Woodruff

13
ON OUR CONSCIENCE

A tree is a tree. How many more redwoods do you want to look at?
If you have seen one, you've seen them all.
Ronald Reagan

By the beginning of the twentieth century, man had finally come round in a full circle. At last he had learned that he cannot subdue the highly resilient flora but must work with it. Plants return again and again like a well trained corps of infantry, with new settlement patterns based on themes which started at the end of the Great Ice Age. Today the landscape is an uneasy balance between our material needs and nature, with the full knowledge that if man stops cultivating or grazing the land, the plant army will return. The land will become woodlands again in the valleys and on their lower slopes. Scrubland will appear in places as well as on the upper valley slopes and moorland will return to the tops of hills and mountains—a repeat of what happened after the Black Death.

The twentieth century was to see two world wars and a huge increase in the population, along with vastly increased road and railway systems. Curiously, the local impact of Iron Age Man on the wildwood 3,000 BP, was far more devastating than the events which were to come in the twentieth century. Then the plough was developed from the discovery of how to use iron, which became the tractor and the combine harvester of later centuries. Care must be taken in any review of events in this twentieth century, for being too close to the immediate past puts changes and losses out of perspective.

In the past, periods of social calm and increased literacy have increased the knowledge of the flora and the changes which were taking place at that time. The beginning of the twentieth century was calm enough, but the social fabric of society was beginning to crack as the huge landowners became increasingly short of money. The road and rail networks continued to expand, creating new habitats such as embankments, verges, sidings and cuttings, at the same time as they unfortunately destroyed old ones. Quarries appeared to provide stone for the new roads and railways. All these areas were prime sites for native and alien plants alike, to establish a home. No wonder that pineapple mayweed, Oxford ragwort, monkey flower, and buddleia among many more, were able to become so well established. It was the world's 'Great Rock Garden' all over again—the only difference was that it was man-made. Sadly, what we did not recognize was that some plants could not move their home and were lost for ever.

Gravel and clay pits, also man-made, are on areas of sand, small stones, and gravel left behind by the glaciers as they finally retreated for the last time at the end of the Great Ice Age. When they were dug out for the gravel and sand which they contained, the holes left behind filled up with fresh water. This provided homes for fresh water plants threatened by the increasing pollution of the rivers and the man-made alterations in river courses. Some were developed as reservoirs of clean water for urban supplies, while still providing a habitat for water plants which had settled there earlier.

Many of these sites can be seen from a train travelling on the Oxford to London Paddington route, as well as from stretches of the M4 going east to London from Swindon. These are large examples of something which has been happening for centuries. Many farms have small ponds in their fields which are marl pits. These have been dug over time, for they are a source of clay to be used on the land. Unfortunately many of these have now been drained so more plant refuges have been lost. Pennyroyal is probably a casualty of such drainage.

By the mid 1930s there had been some attempts to investigate what changes had and still were occurring. The fens were being drained again and plants such as fen ragwort had been lost. The military orchid had gone, due to lack of pollinators in a changing climate or the need to make saloop. On the danger list at that time were several more orchids such as the monkey and lady's slipper orchid, red helleborine and lady's tresses. Primroses and daffodils were at risk, again, due to the over-collecting for private gardens, for then, as now, wild flowers, which came free, were expected to grow in our gardens. In 1935 A.J. Wilmot sarcastically remarked that the one species which was on the increase between 1914-1918 was the 'alpine gardener'. He felt that this was a menace; a force for the extinction of the already naturally sparse, upland flora. It had ransacked the mountains for their rarities, the majority of which did not survive the change of habitat, from the mountain to the back garden.

Other reasons for the decline of some of our native flowers were cleaner crop seeds and school nature study lessons! Agriculture was slowly becoming more scientific and the increased use of herbicides and insecticides, meant a higher yield and cleaner crops. The result was that the golden corn marigold, the bright red poppies, azure blue cornflowers, and the purple corn cockle gradually became the rarities that they are today. In the schools, the new Nature Study lessons meant over-collecting of conspicuous flowers for classwork on a regular basis. The numerous flower collections must have reduced the distribution of many flowers, but how many plant species were completely lost is impossible to tell.

On the other hand, then as now, a 'hands on' approach in science has always been the best teaching method. This would have been at odds with the 'do not pick' instructions, so that left the early teachers in a quandary. At least by the 1980s, the National Curriculum for Science had removed that hazard, though earlier conservation attitudes of the 1970s did have some effect. Pressed flower collections had been banned in many primary and secondary schools before the mid 1980s.

The increased trade and improved transport systems brought in aliens through the dockyard gates in the poorly cleaned grain, imported chicken feed, and the wool destined to become shoddy. American willowherb and Mexican fleabane arrived and their light seeds were transported around the country on the air currents made by the increasing number of trains, lorries and cars. The improvements in road services also account for the spread of the Asia Minor – Caucasian slender speedwell. It had arrived in the nineteenth century rock gardens, a plant with vigorous growth. By the 1920s it had been realized that its phenomenal spread was due to pieces falling off the rubbish carts on the way to the town dump. It only took one small piece of a stem or root to produce a new plant. Along with another willow herb from New Zealand, these four plants made their homes on the newly created walls, railway lines, and derelict sites and are still here today.

After the Second World War the twin demands of communication and food on their own probably accounted for our largest habitat destruction. Iron Age Man, when he developed the plough, changed the landscape forever by removing the wildwood to make fields. The Romans had had an even greater effect producing a level of environmental change which even the worst excesses of the Industrial Revolution did not match. The Enclosure Acts destroyed more of the countryside but gave us verges and hedges. These were new habitats, refuges for herbaceous plants, shrubs and trees alongside the new roads.

The building of the Roman roads on their agger banks, even with the local objections, was a calm process compared with the planning and protests before the building of the present motorway networks. The long tussle over the motorways at Twyford Down and Newbury showed that there is a heightened awareness of the past damage to the countryside and a need to conserve what is left.

These linear Roman-like highways have a beneficial side, they too have banks. All these new banks were originally sown with an average of four native species of grass and white clover. They have been planted in various patterns with native trees such as hawthorn, hazel, blackthorn, willow, elder, and dogwood, according to the soil type and bank aspect. Since there is no spraying, with mowing for only the first two years to discourage dock and thistles, colonization by that elite corps of plant infantry was able to take place. It is impossible to exaggerate the importance of these new man-made habitats in the creation of plant refuges.

On the M1 between Hendon and Leeds, the Nature Conservancy recorded nearly four hundred species of plants on the newly-made banks by 1970. Apart from primroses, cowslips, hemlock, and woad already mentioned, in the late 1990s the banks of the M5 have thriving colonies of wood anemones, soapwort, Oxford ragwort, and bluebells. The banks of the M50 near Newent and Ross-on-Wye are extensive refuges for the wild daffodil, while the M4 in South Wales is a huge refuge for primroses. It is like a second coming, Roman roads to motorways.

Finally, not only have the motorway planners had to listen to the plant lobby, the District and Local Authority Highways Departments have been pressured as well. Some verges are now clearly marked sites of interest and are not cut, mown, or sprayed. This was part of a reaction to the wholesale destruction of verges at the same time that the motorway development increased. These man-made features are as important today as they were during the enclosure period, as refuges for rare and common plants alike.

The effects of railway expansion have already been described in detail and the process continued into the twentieth century. Here though, nature has had the last word. In the end the balance between 'need and nature' tipped in favour of nature as the Beeching Plan cut deeply into the railway network. Deserted tracks have become nature reserves, rich in adventives, who had fallen off the rolling stock, joining the calcicoles growing on the clinker left behind when the rails and sleepers were removed. It looks so real that it seems as if the lines had never been there and steam engines are phantoms of the mind.

The changes in both farming and forestry practices after 1945 were essential to feed a rapidly increasing population. By 1950 agriculture was more prosperous and the uneasy balance between 'need and nature' tipped in favour of need. There was a headlong rush to destroy woodland, uproot hedges, straighten streams or drain them, pollute streams and rivers with excess fertilizer and pesticides. This was extreme vandalism on a huge scale. The effect of increasingly mechanized farming and specialized agriculture was to produce areas of monoculture. These can be seen in the 'prairie' landscape of East Anglia and the Hampshire Downs. Man holds back some of the plants by early ploughing, but it favours the early colonizers of open soil. Plants like the plantains and dock, here since the Great Ice Age, have come back from the edges again. Now they are the weeds of modern farming.

This time of vandalism is called the 'Locust years' by Oliver Rackham, who extended the changes in the countryside to include churchyards. 'God's Acre' became incredibly orderly, instead of wild flowers around the plots there were metres of well fertilized and 'pesticided' grass . Fortunately the wind of change has blown over those sterile places, for today not all churchyards are this type of desert. The author's village churchyard has a large area managed as a hay meadow. As a result, it has a varied meadow flora including plenty of false oxslip in the spring, a botanically rich and peaceful example of 'God's Acre'. Probably many village churchyards are just the same, for there was never sufficient time or money to keep them as tidy as Oliver Rackham suggested.

Vandalism was easy to see in the work of the Forestry Commission after 1945, for more woodland was lost in the twentieth century than in the previous four hundred years. The Commission destroyed many woodlands to create farmland, although many of the sites had been rejected during the 'Dig for Victory' campaign of the second World War. The plantations, which were eventually planted in some areas, could never re-create what had been lost, especially when they were planted on cleared heathland which had never been wooded. It is impossible to re-create old woodland because it was never ploughed. The characteristic associations of plants like bluebell, wood sorrel, dog violet, wood anemone, yellow archangel, and herb paris will not naturally re-colonize. They cannot be artificially planted, so re-creation is never an option. Re-colonization will occur but it will always be different.

Stinging Nettle

14
NEW PLACES TO HIDE

We can help make the world safe for diversity.
John F. Kennedy

All living things need a refuge, a place to hide, to which they can escape in the face of danger. Plants have retreated to their refuges, to return again and again when the danger was passed. However in the time after 1945, the vandalism in agriculture and forestry destroyed plant habitats and their refuges alike, leaving few places to which plants could retreat. This was a sad period in our plant history and yet the robust nature of our flora was to carry it through this devastation, towards a brighter and more prosperous time in the late twentieth century.

The flora responded to the need for motorways by colonizing the banks of the carriageways and slip roads of the new motorway network. It had returned to claim its earlier territory as the railway lines were closed. However these places were insufficient for the huge numbers of displaced refugee plants whose habitats and refuges had been completely destroyed. This was a problem which could not be solved by these two sites alone.

Curiously, in spite of man's best efforts, some of the land never lost its natural flora. It returned when a man-made disaster accidentally released it. In 1954 the rabbit myxomatosis epidemic proved to be a massacre on the level of the Black Death. Myxomatosis had been introduced by man in an attempt to control a growing rabbit population by means other than shooting. It meant that for the first time since their Norman introduction, there were insufficient rabbits to eat the new sapling growth and nibble the spring grass. The result was that old pastures became woodlands again. The downland bloomed with orchids and its turf became rejuvenated with a mass of new species of flowers. When the new breed of myxomatosis-resistant rabbit appeared, a few areas of land were returned to their original downland state. Fortunately however, some of this re-creation has been permanent—a bonus at a time of gloom.

An even more astounding story of the natives' return has been recorded in Cornwall. Here a conservation-minded farming family has recreated the grazing pastures on land facing the Atlantic coast. They uprooted the gorse and bracken—no mean task—from the headland pastures. With the 'help' of their herd of highland cattle and moorland ponies, they managed the land as grazing pasture from early autumn onwards much as it had been used in previous centuries. The flowers of the pasture, now released from the overpowering gorse and bracken reappeared the following spring. There were violets, orchids and wild thyme, among the many downland plants which came out to greet the farmer, who in effect had put back the clock.

By the 1970s vandalism in agriculture and forestry had reached a peak and a more caring and informed attitude began to appear. Scientific working practices, rather than a form of the 'slash and burn' system, were employed. The active management of ancient meadows and woodlands appeared, hedges and verges were protected and cut carefully, and the flora began to return as it had so many times before. As well as the new wildlife trusts, other organizations seemed to become aware of their responsibilities. Hillingdon Council began to manage the ancient Ruislip Woods along the old lines of coppicing the hornbeam on a twenty year cycle, using shire horses to transport the cut wood. Burnham Beeches has its beech woodland regularly coppiced. Local councils began to designate areas for conservation and it suddenly seemed that there was a more enlightened attitude to wildlife as a whole.

In spite of these accounts of the rejuvenation and conservation of our land, it was very tempting, and perhaps realistic, to be gloomy about the state and viability of our flora post 1945. Where was it going to find places of safety, from which it could return or start again? Curiously this was an area of positive development. The increasing number of man-made sites gradually became not only refuges but new habitats, for our diverse and resilient flora to conquer and colonize successfully.

A curious and unexpected place to find a refuge or even habitat development is an airport. The vast expanses of runways with grass in between them are extremely safe, for no one will trespass there, when those great, longhaul jets are moving around. Those airport perimeters with woodland and meadows are also good refuges. Since noise is not a factor which disturbs plant growth, as proximity to motorways has shown, airports are potentially good plant refuges, if one can get there alive!

Access is also a factor when assessing the potential of refuse dumps either covered over or still in use. Sewage farms are well-known refuges for they, too, are relatively undisturbed and easier of access. Many ruderals have been found here as well as herbaceous plants, shrubs and trees. A list from here would include fat hen, several willowherbs, ox-eye daisy, mugwort, hazel, elder, and alder. These would all provide a pool of plants ready to move out when there was space on the land round about.

Gravel and clay pits are safe refuges as long as they are not over-used for leisure activities. Quarries, even before blasting and the collection of stone is finished, are excellent refuges, especially if they contain water. After all, this is just like the landscape after the Great Ice Age. Some of the older sites, where quarrying has ceased, like those on the Malvern Hills and the Isle of Portland, have rich colonies of plants in the loose rocks left behind, as well as on the access roads. Some quarries have become nature reserves like Wingate Quarry in County Durham. At Hay Wood near Walsall the mine workings, with their associated fresh water canals, have been developed as a nature reserve. All these sites have great potential for refuge and habitat development, since they are of no value in any future type of industrial or building scheme. Wetlands produced from mining subsidence, such as ings or flashes, are also valuable as they have no potential for redevelopment either.

Surprisingly the urban area is a rich if a little risky place for plant refuges. These are found around car parks, on road sides and on walls. Plants there can often reflect the early use of the land as well as what has come in from outside the town. An example of a plant relic of early use would be vipers bugloss, a cure for lumbago, growing on a wall near an old monastery physic garden. Wallflowers, whose seeds were blown in from urban gardens, often grow well on old walls and start to shelter a colony of other plants of various origins.

The sides of the wall can provide places for plants like common fumitory to grow if the mortar is loose, or a little rough. The wall bases, like hedgerow bottoms, are well protected places for plant growth, nourished by the detritus blown in the wind, which eventually lands at pavement or road level. Plants such as tutsan, hairy bittercress, yellow rocket and pineapple mayweed thrive in these conditions. Unfortunately, these places are vulnerable to pollution and man's urge to clean walls and sweep the pavements. In fact, walls are the best refuges in an urban area, especially if they are undisturbed.

Other excellent plant refuges, which are unfortunately at risk, are our

13. *God's Acre at Alfrick managed as a meadow.*

14. Ragwort

15. Polyanthus x Cowslip

16. *A landscape mosaic.*

rivers and coasts. As factories and new homes are built around rivers, the channels are altered and access to the hinterland destroys more of the landscape. On the coasts salt marshes and sand dunes are being lost to various types of development, as well as being trampled in the rush for recreational space. Dunes like those at Ynyslas in Dyfed with their carefully planned and raised walkways are unfortunately rare. Sadly even these refuges are at risk from the 'Nibble Factor'.

There is always a need for the land to be used for something else. These 'nibbles' can include land for widening the road, draining and enlarging ditches with a loss of verges, or for larger items such as car parks. The development of new leisure areas such as golf courses, housing estates, and large supermarkets with their essential access roads, all remove areas of land which decreases the total land surface for plant refuges and habitats. The risks are further increased by the planting of non-native trees such as cypress, horse chestnut, and balsam poplar, around the car parks and access points. If native trees cannot grow there, then it was as nature intended. These non-native trees sour the soil and often darken the landscape. The construction of golf courses, often described as 'natural leisure areas', totally changes the landscape by removing much of the natural vegetation, changing the drainage and finally by replanting with either non-local or non-native trees. This totally upsets the natural, ecological balance of the area.

All habitats and refuges, natural or man-made, are being subjected to a new and insidious attack. Around large developments such as shopping centres, motorway service areas and so-called leisure centres, contractors are sowing plant seed mixtures to naturalize the areas. Unfortunately many of them contain seeds of continental or artificial origin. When pollination takes place between these newly-sown plants and the local ones, the resulting plants are neither the newly planted ones or the local, native species. A good example of this is a new species of birdsfoot trefoil, showing considerable hybrid vigour. It will eventually out-compete the local species, which unfortunately will be lost. Other plants already at risk in this way are cornflower and corncockle, whose status gives cause for concern as rare and rapidly disappearing species. The new species are being planted for their natural appearance, colour, and hybrid vigour.

Other destroyers of plant communities are the introduced animals, which are non-native and not part of the plant-animal balance. The rabbits have already been mentioned in this context and deer, some of which were brought by the Normans, can be included as well. The muntjac deer, fallow deer, and Chinese water deer all have a disastrous effect on the local plant life if they are not contained. Muntjac deer achieved local notoriety by escaping from the park at Longleat to feast on the bluebells in a local woodland. If they are kept here, then they must be controlled or again there may be long term, disastrous effects. The effect of these browsing mammals is to deflect, often permanently, the natural plant succession.

The present range of polyanthus plants have their origins in the sixteenth century, when natural 'sports' or 'rogues' are first recorded. Gerard records a double variety of the cowslip often called 'hose-in-hose', possibly a cross with the primrose or oxlip. There would appear to have been two primrose 'sports', one of which was green and the other white or very pale yellow. Since the seventeenth century these plants have been cross-bred to produce garden polyanthus. Unfortunately, when the garden polyanthus crosses naturally with the cowslip there is another strain of polyanthus × cowslip. These crosses are common and often very brightly coloured. If these become fertile, then man has provided another coloured thread to the multi-coloured strands of our floral history, for he will have added to the natural gene pool . The natural crosses and mutations are part of plant evolution, but the addition of artificial or European seed will diminish our flora, as the local wild flowers, totally out-competed, will disappear completely.

Corn Cockle

THE COMMUNITY FOREST PROGRAMME
IN ENGLAND

Great North Forest
• SUNDERLAND

• MIDDLESBOROUGH
Cleveland Forest

South Yorkshire Forest
Red Rose Forest
• SHEFFIELD
LIVERPOOL •
The Greenwood
The Mersey Forest • NOTTINGHAM

Forest of Mercia

• BIRMINGHAM
Marston Vale • BEDFORD
• ST. ALBAN'S
Watling Chase
Bristol-Avon Forest Great Western Forest Thames Chase
• SWINDON
LONDON
• BRISTOL

15
THE NEXT STEP

And out of the ground made the Lord God to grow every tree that is pleasant in the sight and good for food; the tree of life also in the midst of the garden and the tree of good and evil.
Genesis

When the writers of the Old Testament were describing the Creation for the Children of Israel, they perhaps instinctively understood man's love of trees. Today the same thing has happened, for man has turned to trees as a starting point for a conscious effort to manage and save his flora. Can this seen in the popular activity of going for a walk in the woods? It is a marvellous alternative to the town or city parks with their ordered way of life. There are fewer rules and regulations here, for we can roam more freely, roll in the leaves, climb some of the trees, and play mini war games without being told about the noise. In the English make-up there seems to lurk a wood. A Mori poll for Shell showed that nine out of ten British people visit woods regularly, one in six claim to have planted trees, one in four has fallen out of one, and one in three has walked into one. The author comes into all these categories and has the scars to prove it!

This love affair with trees still exists, for half the trees in Britain are in cities. Trees like the London plane and common lime cool the air in summer, warm it in winter, absorb pollution and act as sound baffles. Hospitals report that patients recover better when they can see trees. They are now planted on old factory sites like the one near Mosely in Liverpool. Regretfully street trees are at risk, not just from vandals, but from lack of water. This is due to the digging up of their water-absorbing roots by the water boards laying mains, together with the cable-laying work of the telephone companies and the cable networks.

This land was once covered by the Wildwood, but around 6,000 BP man started to clear it. Since then there has been little conscious effort to look at what damage has been done. As the millennium approaches, man realizes that he cannot recreate what has been removed. Fortunately it is possible to replant the land and create new woodlands. If native trees are used in this planting, secondary woodland will appear with new ecosystems, providing the necessary niches and refuges for the plants to colonize. So, when we 'go down to the woods today' there is a security in knowing that someone cares about trees, which may be the starting point for a new age of plant consciousness.

On the Cheviot Hills in the north of England, there is a cunning plan afoot to re-create the original Border Country. Today it is a bare sheep-walk landscape with some commercial plantations of sitka spruce. In the days of the Border Raiders, war was fought in both directions over the ownership of land, cattle, sheep, or women. There were broad-leaved trees such as birch, ash and oak with alder and willow by the streams. As the seventeenth and eighteenth centuries passed and life became more peaceful, the land was slowly cleared for farming. The Forest Design Plan of the Forest Enterprise—once part of the Forestry Commission—is to replant several acres of native broad leaved trees. This will increase species diversity in the area, as the developing, secondary woodlands are colonized by flora finding new refuges. In time there should be a new look to large areas of the Border landscape.

Coming south into the Midlands, there will be the National Forest which will transform almost two hundred square miles of countryside. It will extend across Derbyshire, Leicestershire and Staffordshire, blending in the ancient woodlands with new planting around the farms, villages and towns. Initially this idea was slow to start, in spite of generous funding for the purchase of a wide range of native broad leaved trees. The generally accepted definition of a forest is 'a wooded area which will meet several purposes'. The National Forest will fit this definition since it will provide leisure and learning resources as well as greening a landscape which was first cleared by Iron Age Man. The Enclosure Acts completed the clearance leaving the Midlands the least wooded part of England.

Woodland management in the south of England has a different problem highlighted by the Timber Growers Association. It points out that many woodlands in this area have been subjected to a 'boom and bust' treatment, seen clearly at the end of the nineteenth century. Today there are many small woodlands, especially beech ones, which have no saleable timber but are still in need of management. The lack of cash-flow is a problem haunting their owners, who can see that in one good storm, a great deal of damage may occur which will be costly to repair. An alternative at this point is to clear-fell the woodland, which is not a popular option. The 'nibble' factor is often present for these non-productive woodlands, which makes the present range of grants for woodland development even more important.

On a more positive note there is an exciting project in hand, involving the Sussex Wildlife Trust's small area of wildwood. The Andreswald project will attempt to link up this woodland with a second one, by purchasing some of the low-quality, agricultural land in between to create a wildlife corridor. Once the woodland is established it is planned to introduce wild animals, not the bears and aurochs of the original forest, but some present day equivalents—pigs, ponies, and cows have been mentioned. The idea is to recreate the clearings and trackways of an ancient woodland, manuring it naturally and hopefully creating refuges for plants and animals alike.

No account of activity in the country's woodland would be complete without the Community Forest programme. Aided by the Countryside Commission, the Forestry Commission and, at the last count, sixty-two local authorities, the Community Forests shown on page 106 have spread from Bristol Avon in the south to the Great North Forest in Northumberland and County Durham. Some were planted on disused agricultural land around towns and cities, others are on areas of industrial dereliction including town dumps. Swindon's Shaw Tip has six thousand saplings of a variety of native broad leaved trees. Apart from a huge range of leisure activities from photography, plant and animal identification to riding and walking, these secondary woodlands will provide more refuges for some of our beleaguered plants.

Wildlife corridors will eventually connect many of these woodland refuges. They may be very small, such as the banks of a stream, canal or river but they create a spider's web effect over the area, allowing the movement of plant and animal species to take place. Set-aside land has been of great value here, as have the disused railway lines. Careful future planning by local authorities and farmers alike will ensure that these corridors are increased and expanded.

A different effort is being made to save nearly three hundred plants on the verge of extinction. The Royal Botanic Gardens at Kew have applied to the Millennium Commission for £41 million to develop a seed bank at Wakeham Place. Parts of the country have recently recorded the loss of some native plants, Worcestershire has lost corncockle, Leicestershire has lost snake's head fritillary, and yellow-wort has disappeared from Cumbria. The changing use of the landscape has created these crises.

The heathland areas of Dorset had been decreasing in size for at least two centuries, due to the increased on-site building and lack of management. Lowland heathland is part of the post Ice Age building site, the area where sand and gravel were dumped by the retreating ice. It developed its own unique ecosystem, a complex gorse and heather relationship on soil low in nutrients. Land management had been by grazing by ponies and sheep, without which the invasive hawthorn, rhododendron and birch would take over producing a scrubland.

Today there is at least ten percent more heathland as a result of an EU grant for less well-off, local land owners. This money has been used to clear some of the scrubland and re-introduce grazing. Slowly the heathland ecosystem, the inspiration for Thomas Hardy's Egdon Heath, is returning with its unique mix of plants and animals—refuges for all concerned, including a rare sand-lizard— *Lacerta agilis agilis.*

It has been said before that the British are a nation of shopkeepers, a profession which needs regular stock-taking if it is to be efficient. Regular stocktaking is essential if the frequency and distribution of our flora is to be monitored. This has been the work of the hard working and often un-acknowledged local conservation groups and trusts. Frequently as volunteers, they struggle against a sea of paper and miles of red tape from both local authorities and central government. For it is these groups who buy and manage land as reserves, organize plant and animal counts and record the data, while keeping a watching brief over events on their patch. These events can be as diverse as new housing developments, road widening, or changes in land use which will affect the wildlife in their areas.

Plans to control the development of the countryside for the next century are essential if the existing flora is to survive. There are already skills to manage properly the mosaic landscape of today and the needs and uses of hedgerows and fields are now better understood. The value of copses, woodlands and forests are also appreciated. The lowlands around rivers and at the sea shore are more carefully managed than before, as their benefits become obvious. Unfortunately, as is often the case, there are unforeseen flaws which could alter these plans considerably.

One of them is natural hybridization. This is the crossing between two related species, which has always been present. Mention has been made of the false oxlip from the primrose × cowslip cross, as well as the cowslip × polyanthus cross, producing brightly coloured cowslip plants. This process is continually occuring, providing a changing plant population in which natural selection will take place. This leads to the much-discussed spectre of man-made genetically modified plants, otherwise known as GMPs; twentieth century genies emerging from a laboratory bottle, not by natural selection.

Their older brothers, the hybrid crop plants, have been with us since in the ancient civilizations early man started to select the strains of cereals which produced the greatest harvest. As scientific skills have advanced and become more sophisticated, the majority of his food plants have been bred for such factors as increased yield, wind resistance, and poor drainage. The GM crops take this development one stage further. By manipulating their genetic make-up, they can be made insect-resistant. In others words, insects do not eat them; an effect incompletely produced in the past by spraying. It is here that there is considerable cause for concern.

If insects cannot get limited food supplies from the crops, they will eat other plants in the ecosystem, with potentially damaging results. The hunt for food, which includes nectar, will change direction and the pollination routes will be interrupted. This will have the effect on the wider ecosystem by altering the level of seed production in the autumn, so affecting the number of plants the following year. The ripple effect on the countryside may be enormous and has yet to be explored, as has the effect on the human metabolism of these genetically modified foods. It is unfortunate that many of these GM crops have great advantages for the developing world, where insect damage frequently causes famine and death.

Another problem in the management of the countryside is the use of reed and willow beds. These are frequently planted to 'clean' water from septic tanks and other effluents from a wide variety of sources. These plants absorb toxic elements as well as domestic and farm waste, so that the water leaving the beds is relatively clean and useable. This begs the question; when these plants die are the undesirable elements or compounds released into the ecosystem? If the answer is in the affirmative what steps need to be taken to keep the levels of pollution at an acceptably low level?

As well as these problems, yet another very subtle and effective genie has escaped from his bottle, never to be pushed back inside. He is called

'Global Warming'. Because man has burned fossil fuels and forests producing excessive amounts of carbon dioxide, this has prevented the loss of heat from the earth out into space. As a result, over the last one hundred years there has been an average rise in temperature of 0.5° C. Further similar temperature rises are forecast for the next century.

Here in England there have been earlier climatic changes, both large and small, from the warm Bronze Age to the cold winters of the seventeenth century. World climatic changes were recorded when Tambora erupted in 1815, which in 1816 produced no summer or harvest over the whole of the northern hemisphere. The level of solid material in the air blocked out any sunlight of value and is captured for eternity by the artist Turner in the lurid red sunsets of his landscape paintings.

So far, all these changes appear to have had minimal effects on the flora, but what will happen to it in a increasingly drier, warmer England? Some southern shore lines may not be backed by a 'green and pleasant land' but will appear dry and dusty. The soft greens and pink blossoms of springtime in the 'Garden of England' would be no more, for the soil would be too dry and the climate too warm. It outrages many people to think that the characteristic southern downland and colourful, leafy lanes of southern England may look like a desert. On the positive side, there could be orange orchards, hedgerows of bougainvillea, and fields of sweet peppers and evening primrose. The latter is already here thriving in the warmer weather, though not a yet cash crop. A sad blow would be the loss of famous pieces of the English landscape such as Egdon Heath and the estuaries of the lower Thames immortalised in Charles Dicken's *Great Expectations*.

The rise in sea level may mean more sea defences being built, cutting the carefully created plant and animal corridors, so preventing migration away from the rising waters. Towns, cities, railways, and roads, will all halt the movement north of plants which cannot survive the rise in the average, annual temperature. Curiously this is not new, for it happened at the end of the Great Ice Age when the ice and rubble left behind blocked the route north. This book has attempted to describe what arrived here as a result. Eventually, global warming may create new habitats in southern England, which will develop thriving colonies of continental species, increasing the natural diversity of our flora. This is not a doom and gloom scenario, for it has happened before—it is the start of something new and exciting!

Plants were here long before mankind walked onto the land which was to become England. Their mysteriously silent and majestic world has subsequently fed, housed, and clothed us, as well as provided medicine and plants for pleasure in the garden. By changing the natural landscape to suit agriculture and commerce, many of the early habitats and their unique assemblages of plants have been lost, never to return. This has been the inevitable price of progress with ignorance. It is important to take into the next century an awareness of what happened in the past and use that knowledge to monitor future events. Only then will the flora be protected from man's excesses. Careful management of the present habitats, with their unique assemblages of plants, is the way forward into the next century. This will ensure that our flora, native and non-native, can grow in safety, so that its diverse history may continue to delight future generations of plant lovers. This would be the price of progress with knowledge .

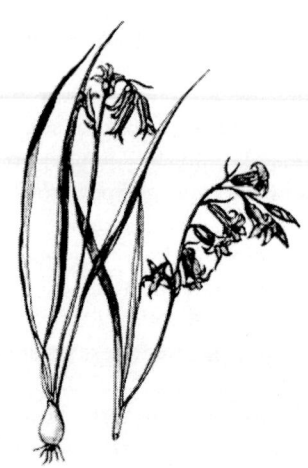

Bluebell

EPILOGUE

*The flora is not ours to own, but ours to hold in trust.
As a unique inheritance, it is to be cared for
as an investment for future generations.*

APPENDIX A
English plants in North America

- Burdock
- Bugloss, Viper's
- Celandine, Greater
- Chickweed
- Cinquefoil
- Clover, red.
  - white
- Comfrey
- Couch grass
- Cress, Thanet
- Dandelion
- Dock, curled
- Dyer's greenweed
- Fat hen
- Groundsel
- Heather
- Hellebore, green
- Hemp
- Houseleek
- Ivy, ground
- Knotgrass
- Marjoram
- Mayweed, stinking
- Nettle, stinging
- Pennyroyal
- Plantain, Greater
- Shepherd's Purse
- Soapwort
- Star of Bethlehem
- Storksbill, musk
- Tansy
- Thistle, creeping
- Toadflax

APPENDIX B
English Plants in Australia

- Blackberry
- Briar, sweet
- Broom
- Buttercup, creeping
- Charlock
- Chickory
- Clover, red
  - white
- Dandelion
- Dock, curled
- Fat hen
- Fennel
- Fescue, rat tail
- Fleabane, Canadian
- Golden rod
- Gorse
- Grass, Johnson
  - marram
  - meadow
  - perennial rye
- Grass, quaking
  - sweet vernal
- Hemlock
- Ivy
- Knotgrass
- Nightshade, black
- Pellitory-of-the-wall
- Periwinkle
- Pimpernel, scarlet
- Plantain, ribwort
- Reed, giant
- Shepherd's purse
- Spurge, petty
- St John's wort
- Sowthistle
- Thistle, creeping
  - spear
- Vetch, common
- Yarrow

APPENDIX C
English Plants in New Zealand

Bindweed, field
Blackberry
Bugloss, viper's
Charlock
Cress, Thanet
Dandelion
Dock, curled
Fat hen
Fleabane, Canadian
Grass, rye
Grass, meadow
Knotgrass
Plantain, greater
Ragwort
Shepherd's purse
Sowthistle
Thistle, creeping
    smooth
Yarrow

APPENDIX D
Some Common Medicinal Plants

Plant	Use
Balm	Tonic wine
Basil	Sedative: indigestion remedy
Betony	Tonic wine
Broom	Diuretic
Bugloss, viper's	Anti-depressant: lumbago: snake antidote
Celandine, greater	Eye complaints
Coltsfoot	Chest and throat infections
Comfrey	Poultices for broken bones
Dandelion	Diuretic
Elder	Coughs and colds: skin cleaner: eye lotion
Eyebright	Eye wash
Feverfew	Headaches
Foxglove	Diuretic
Hemlock	Painkiller
Henbane	Painkiller
Ivy, ground	Liver complaints
Lavender	Headaches: insomnia: indigestion
Lords and Ladies	Convalescent drink
Lungwort	Lung complaints
Marjoram	Cure-all, from headaches to bladder problems
Marigold, pot	Antiseptic: skin inflammation: varicose veins
Mugwort	Cure-all from stomach upsets to insomnia
Nettle, stinging	Gout, asthma
Orchids, green-winged / early purple / military	Convalescent drinks
Rosemary	Rheumatism: heart complaints: antiseptic
Rue	Eye complaints: headaches: anti-infection: holy water
Sage	Tonic: insect repellent
Sneezewort	Toothache
Tansy	Prevention of miscarriage
Thyme	Nausea: headaches
Wormwood	Tonic: digestive upsets: anti-pinworm

APPENDIX E
Plants about the House

Ash	Pick-handles
Bedstraw, Lady's	Mattress filler
Beech	Furniture: chopping blocks
Birch	Houses: furniture: utensils
Bluebell	Glue
Broom	Cleaning tools
Butterbur	Cleaning
Burdock	Insulated wrapping for food.
Butcher's Broom	Cleaning
Coltsfoot	Beer: jelly: wine: tinder boxes.
Couch grass	Coffee substitute.
Cranberries	Silver cleaner: preserves.
Elm	Flooring
Hazel	Walls: fencing
Heather	Bedding
Lime	Carving
Hops	Pillows: clears beer
Horsetails	Metal cleaner
Iris	Black ink
Lords and Ladies	Starch for clothing: saloop
Marsh marigold	Fertility flower for cattle
Mugwort	Moth repellent: flavouring for beer
Nettle, stinging	Tea: beer: fibre for cloth: frog repellent
Nuts	Polishes
Oak	House and window frames: furniture
Oats	Flies for fishing
Old Man's Beard	Tobacco substitute: pipes from stems
Rue	Disease: insect repellent: sanctified water
Soapwort	Washing substance
Sorrel	Removes rust marks: flavours beer
Tansy	Food colouring: insect and mouse repellent
Teazle	Combing out wool
Tormentil	Tanning leather
Walnut	Decorative pieces

APPENDIX F
Aromatic Plants about the House

Basil
Balm
Chamomile
Fennel
Flag, sweet
Fleabane, common
Hop
Lavender

Marjoram
Meadow sweet
Melilot
Mint
Pennyroyal
Savory, winter
Tansy
Tutsan
Woodruff

APPENDIX G
Plants Used for Dyes

Black:
    Gipsywort
    Iris, yellow or flag

Purple:
    Bilberry
    Oak bark

Brown:
    Bracken leaves
    Lichens
    Oak: sawdust and oak apples

Yellow:
    Agrimony.
    Broom flowers.
    Dyer's greenweed
    Fennugreek
    Lady's Bedstraw: leaves and stems
    Marsh marigold: from plant, with alum
    Weld.

Red:
    Alkanet
    Fat hen
    Lady's Bedstraw: from the root, with alum
    Madder
    Privet berries
    Tormentil roots

Blue:
    Privet berries.
    Woad

Green:
    Bracken: from young shoots
    Privet: from berries, with alum
    Reed flowers

APPENDIX H
Plant Index Key

A = Adventive   C = Colonist
D = Denizen   N = Native
G = Garden flowers which have not escaped into the flora.

PLANT INDEX

Common name	Pages	Scientific name	Code
aconite, winter	54	*Eranthis hyemalis*	D
alder	5 11 81 102 107	*Alnus glutinosa*	N
alehoof or ivy, ground	84	*Glechoma hederacea*	N
alexander	28 35	*Smyrnium olusatrum*	D
alkanet	43	*Anchusa officinalis*	D
angelica, garden	42	*Angelica archangelica*	D
anemone, wood	37 97 99	*Anemone nemorosa*	N
apple, crab	39	*Malus sylvestris*	N
cultivated	2 35 44	*Malus domestica*	G
apricot	42	*Prunus armeniaca*	G
archangel	54 99	*Lamiastrum galeobdolon*	N
arum, wild or lords & ladies	90	*Arum maculatum*	N
ash	11 71 107	*Fraxinus excelsior*	N
mountain or rowan	5 11 36	*Sorbus aucuparia*	N+D
aspen	36	*Populus tremula*	N
avens, mountain	10	*Dryas octopetala*	N
balm	28	*Melissa officinalis*	D
balsam, Himalayan/Indian	79 83	*Impatiens glandulifera*	D
orange	78 80	*Impatiens capensis*	A
small	80	*Impatiens parviflora*	D
barley	1 18 19 33 68	*Hordeum spp.*	–
barrenwort	54	*Epimedium alpinum*	D
basil, sweet	29	*Ocimum basilicum*	G
bartsia, alpine	62	*Bartsia alpina*	N
bay	29	*Laurus nobilis*	G
bear's breeches	29 30	*Acanthus mollis*	D
bedstraw, heath	10 15 36	*Galium saxatile*	N
lady's	10 15 40 90	*Galium verum*	N
beech	5 11 19 46 74 90 101 107	*Fagus sylvatica*	N
bellflower, creeping	41	*Campanula rapunculoides*	D
betony	20 78 91	*Betonica officinalis*	N
bilberry	36	*Vaccinium myrtillus*	N
bindweed, black	17 18 21 34	*Bilderdykia convolvulus*	C
birch	5 7 9 36 81 89 90 107 109	*Betula spp.*	N
bird's-foot trefoil	1 11 15 19 40 72 103	*Lotus corniculatus*	N
birthwort	35	*Aristolochia clematitis*	D
bittercress, hairy	102	*Cardamine hirsuta*	N
blackberry or bramble	9 87	*Rubus fruticosus*	N
blackthorn or sloe	5 39 71 97	*Prunus spinosa*	N
blood-drop emlets	77	*Mimulus luteus*	D
bluebell	9 12 37 44 47 50 97 99 104	*Endymion nonscriptus*	N
bog cotton	18	*Eriophorum spp.*	N
bog myrtle	18	*Myrica gale*	N
borage	2 21 41 43	*Borago officinalis*	D
box	9	*Buxus sempirvirens*	N
bougainvillea	111	*Bougainvillea glabra*	D
bracken	100 101	*Pteridium aquilinum*	N
bridewort, confused	70	*Spiraea x pseudosalicifolia*	D
broom, common	49 87 92	*Cytisus scoparius*	N
butchers	90	*Ruscus aculeatus*	N
bryony, black	39	*Tamus communis*	N
buckwheat	85	*Fagopyrum esculentum*	N
buddleja/buddleia	81 94	*Buddleia davidii*	D
bugloss, viper's	85 102	*Echium vulgare*	N

burnet, salad	75	*Sanguisorba minor*	N
butter and eggs or toadflax, common	85	*Linaria vulgaris*	N
butterbur	77	*Petasites hybridus*	N
buttercup	7 15	*Ranunculus spp.*	N
bulbous	67	*Ranunculus bulbosus*	N
creeping	87	*Ranunculus repens*	N
campion, red	7 12	*Silene dioica*	N
white	16	*Silene alba*	C
cannabis or hemp	34 85 86	*Cannabis sativa*	D+G
caraway	29 30	*Carum carvi*	D+G
carnation	2 30	*Dianthus caryophyllus*	G
cedar	74	*Cedrus spp.*	G
celandine, greater	28 35 78 85	*Chelidonium majus*	D
lesser	12	*Ranunculus ficaria*	N
chamomile, lawn	40	*Chamaemelum nobile (is)*	N
charlock	16 21 28 86 87	*Sinapsis arvensis*	A+C
cherry wild/gean	2 35 39	*Prunus avium, Prunus cerasum*	N+G
chestnut, horse	54 103	*Aesculus hippocastanum*	G
sweet	2 29	*Castanea sativa*	D
chickweed, common	9 43 81 88	*Stellaria media*	N
chicory	87	*Cichorium intybus*	?N
cinquefoil	9 40	*Potentilla spp.*	N
shrubby	20	*Potentilla fructicosa*	N
cistus	58	*Cistus icanus*	G
clary	54	*Salvia spp.*	D
clematis, wild or traveller's joy	9 50	*Clematis vitalba*	N
clover, red	72 81 86	*Trifolium pratense*	N
white or Dutch	56 72 81 87 97	*Trifolium repens*	C
cocklebur, spiny	78	*Xanthium spinosum*	A
codlins and cream	76	*Epilobium hirsutum*	N
coltsfoot	77 81 82 86 91	*Tussilago farfara*	N
columbine	40	*Aquilegia vulgaris*	D
comfrey, common	84	*Symphytum officinale*	N
coriander	29	*Coriandrum sativum*	G
corn cockle	25 33 34 96 103 108	*Agrostemma githago*	C
cornflower	10 41 56 96 103	*Centaurea cyanus*	N+D
corn spurrey	18 25 34	*Spergula arvensis*	C
corydalis, purple	41	*Corydalis solida*	A
yellow	76	*Corydalis lutea*	A
cowslip or paigle	2 12 40 50 51 97 104	*Primula veris*	N
cranesbill, meadow	20	*Geranium pratense*	N
cress, hoary or Thanet cress	77 86 87	*Cardaria draba*	A
crowfoot, water	47	*Ranunculus aquatilis*	N
cypress, swamp	5	*Taxodium distichum*	D
daffodil, wild	12 95 97	*Narcissus pseudonarcissus*	N
daisy	12 15	*Bellis perennis*	N
Michaelmas	59	*Aster novi–belgii*	D
oxeye	40 102	*Leucanthemum vulgare*	N
dandelion	12 15 76 81 84 91	*Taraxacum spp.*	N
deadnettle, red	17	*Lamium purpureum*	A+C
dock	9 12 86 88 97 98	*Rumex spp.*	N
dog's mercury	37 50	*Mercurialis perennis*	N
dogwood	5 12 97	*Cornus sanguinea*	N
Duke of Argyll's tea plant	63	*Lycium halimifolium*	D
eggs and bacon	1	*Lotus corniculatus*	N
elder	97 102	*Sambucus nigra*	N

elder, ground.		28	*Aegopodium podagraria*	D
elm, English	5 10 15 37 47 71		*Ulmus procera*	N
Wych	5 10 11 15 37 39 47 71		*Ulmus glabra*	N
escallonia		80	*Escallonia macrantha*	D
eyebright		61	*Euphrasia officinalis*	N
fat hen	17 18 21 34 86 87 102		*Chenopodium album*	A+C
fennel	29 30 43 87		*Foeniculum vulgare*	D
fescue, rat's tail		87	*Vulpia myuros*	N
feverfew	40 41 42		*Tanacetum parthenium*	D
field gromwell		25	*Lithospermum officinale*	C
field woundwort		25	*Stachys arvensis*	A
flag, sweet		55 91	*Acorus calamus*	D
flax or linseed	1 18 19 33 34		*Linum perenne*	C
fleabane, Canadian	62 81 86 87		*Conyza canadensis*	A
common		91	*Pulicaria dysenterica*	N
Mexican		96	*Erigeron mucronatus*	A
fool's parsley		21	*Aethusa cynapium*	D
forget-me-not, common		7	*Myosotis arvensis*	N
water		76	*Myosotis scorpioides*	N
foxglove	12 44 67 91		*Digitalis purpurea*	N
fritillary, snake's head	44 108		*Fritillaria meleagris*	N
fuchsia		78	*Fuchsia magellanica*	D
fumitory, common	16 102		*Fumaria officinalis*	A+C
yellow, purple			see *Corydalis*	
furze or gorse		36	*Ulex europaeus*	N
gallant soldier	74 81		*Galinsoga parviflora*	A+D
garlic		43	*Allium spp.*	N
gentian	10 20		*Gentiana spp.*	N
gillyflower or carnation		30	*Dianthus caryophyllus*	G
globe flower		20	*Trollius europaeus*	N
goat's beard		53	*Tragopogon pratensis*	N
golden rod	55 56 81 86		*Solidago virgaurea*	?D
Good King Henry		17	*Chenopodium bonus henricus*	A+C
goosefoot	12 81		*Chenopodium spp.*	N
goosegrass or cleaver		94	*Galium aparine*	N
gorse	9 36 66 67 87 100 101 109		*Ulex europaeus*	N
grass, quaking or totter		87	*Briza media*	N
perennial rye		87	*Lolium perenne*	N
sweet vernal		87	*Anthoxanthum odoratum*	N
greenweed, dyer's	84 85 93		*Genista tinctoria*	N
groundsel		81	*Senecio vulgaris*	N
gipsywort		93	*Lycopus europaeus*	N
hawkweeds		59	*Hieracium spp.*	N
hawthorn, common	5 46 71 97 109		*Crataegus monogyna*	N
midland		39	*Crataegus laevigata*	N
hazel	5 11 37 44 89 97 102		*Corylus avellana*	N
heartsease or wild pansy	18 40		*Viola tricolor*	N
heath	20 36		*Erica spp.*	N
heath, Cornish		10	*Erica vagans*	N
heather	20 36 85 109		*Calluna vulgaris*	N
heliotrope, winter		77	*Petasites fragrans*	D
hellebore, green		85	*Helleborus viridis*	N
helleborine, dark red		95	*Epipactis atrorubens*	N
hemlock	27 28 87 91 92 97		*Conium maculatum*	D
western		5	*Tsauga heterophylla*	D
hemp	1 34 85 86		*Cannabis sativa*	D+G

hen and chickens	30 84	*Jovibarba sobolifera*	D
henbane	26 28 91	*Hyoscyamus niger*	D
herb Paris	99	*Paris quadrifolia*	N
hickory	5 6	*Carya cordiformis*	D
hogweed	50	*Heracleum sphondylium*	N
holly	11 52	*Ilex aquifolium*	N
hollyhock	40 41	*Alcea rosea*	G
honesty	56	*Lunaria annua*	D
honeysuckle	9	*Lonicera periclymenum*	N
hop	91	*Humulus lupulus*	N
hornbeam	5 11 46 101	*Carpinus betulus*	N
horseradish	43	*Armoracia rusticana*	D
houseleek	30 84	*Jovibaba sobolifera*	D
hyssop	29	*Hyssopus officinalis*	G
iris, flag or yellow	40 41 44 50	*Iris pseudacorus*	D
ironwort or stinking mayweed or stinking chamomile	84 85	*Anthemis cotula*	N
ivy	86	*Hedera helix*	N
ground or alehoof	84 91	*Glechoma hederacea*	N
Jackanapes-on-horseback	53	*Calendula major polyanthus*	N
Jacob's ladder	10 44	*Polemonium caeruleum*	N
jewelweed or orange balsam	78	*Impatiens capensis*	A
Joseph and Mary or lungwort	42	*Pulmonaria officinalis*	D
juniper	7 9 18	*Juniperus communis*	N
Kew weed or gallant soldier	74 81	*Galinsoga parviflora*	A+D
kingcup or marsh marigold	42 44 47 50	*Caltha palustris*	N
knapweed	15	*Centaurea spp.*	N
knitbone or comfrey	84	*Symphytum officinale*	N
knotgrass or wireweed	9 81 86	*Polygonum aviculare*	N
knotweed, Japanese	78	*Reynoutria japonica*	D
laburnum	54	*Laburnum anagyroides*	G
lady's tresses, creeping	50 95	*Goodyera repens*	N
lavender	30 41	*Lavandula spp.*	G
leek	35 42	*Allium porrum*	G
leopardsbane	56	*Doronicum pardalianches*	D
lilac	58	*Syringa vulgaris*	G
lily, Lenten	44	*Narcissus pseudonarcissus*	N
Madonna	35 41	*Lilium candidum*	G
Martagon	55	*Lilium martagon*	D
lime	11 74	*Tilia spp.*	N
common	106	*Tilia x.europaea*	N
linseed or flax	1 18 19 33 34	*Linum perenne*	C
lobelia	58	*Lobelia urens*	N
lousewort	20 36	*Pedicularis sylvatica*	N
lovage	29	*Levisticum officinale*	G
lungwort	42	*Pulmonaria officinalis*	D
madder, dyer's	43 93	*Rubia tinctorum*	D
mallow	7 54	*Lavatera spp.*	N
mandragora or mandrake	26 28	*Mandragora officinarum*	G
maple, field	5 37 39 47	*Acer campestre*	N
marigold, bur	76	*Bidens tripartita*	N
corn	25 96	*Chrysanthemum segetum*	C
double	53	*Calendula major polyanthus*	N
marsh	42 44 47 50	*Caltha palustris*	N
marigold, pot	29 91	*Calendula officinalis*	G

marjoram	85	*Origanum vulgare*	*N*
mayweed, stinking	84 85	*Anthemis cotula*	*N*
pineapple	94 102	*Matricaria matricarioides*	*N*
meadowsweet	91	*Filipendula ulmaria*	*N*
medlar	35	*Mespilus germanica*	*D*
melilot, common ribbed	51 56 91	*Melilotus officinalis*	*C+D*
small	57	*Melilotus indica*	*C*
milkwort, heath	36	*Polgala serpyllifolia*	*N*
mind-your-own-business	70 76	*Soleirolia soleirolii*	*A*
mint	9 28 29	*Mentha spp.*	*N*
mistletoe	1	*Viscum alba*	*N*
monkey flower	77 78 94	*Mimulus guttatus*	*D*
mugwort	9 81 86 102	*Artemisia vulgaris*	*N*
mullein	44 54	*Verbascum spp.*	*N*
mustard, black	21 28 29	*Brassica nigra*	*D*
white	28 29	*Sinapis alba*	*D*
myrtle, bog	18	*Myrica gale*	*N*
nettle, stinging	1 12 44 81 84 88 91	*Urtica dioica*	*N*
Roman	60	*Urtica pilulifera*	*D*
nightshade, black	12 86	*Solanum nigrum*	*N*
enchanter's	47	*Circaea lutetiana*	*N*
oak, ped.	5 10 34 46 49 52 63 71 73 74 82 89 107	*Quercus robur*	*N*
sessile	5 10 46 71 74 82 89 107	*Quercus petraea*	*N*
oats	33	*Avena sativa*	*D*
old man's beard	9 50	*Clematis vitalba*	*N*
onion	43	*Allium cepa*	*N*
orache	81	*Atriplex patula*	*N*
orchid, bird's nest	54	*Neottia nidus−avis*	*N*
early purple	92	*Orchis mascula*	*N*
green winged	92	*Orchis morio*	*N*
lady's slipper	59 60 95	*Cypripedium calceolus*	*N*
military	92 95	*Orchis militaris*	*N*
monkey	95	*Orchis simia*	*N*
ox tongue, bristly	17	*Picris echioides*	*C*
oxlip	104	*Primula elatior*	*N*
oxlip, false	98 109	*Primula veris x vulgaris*	*N*
paigle or cowslip	2 12 50 51	*Primula veris*	*N*
pansy, field	11 40	*Viola arvensis*	*N*
parsley, cow	72	*Anthriscus sylvestris*	*N*
pear	2 35 39	*Pyrus spp.*	*G*
pellitory of the wall	87	*Parietaria judaica*	*N*
pennycress, field	17 34	*Thlaspi arvense*	*A+C*
pennyroyal/pudding grass	28 29 40 61 84 93 95	*Mentha pulegium*	*D*
peony	35 41	*Paeonia spp.*	*G*
periwinkle, greater	56 86	*Vinca major*	*D*
periwinkle, lesser	9 40 56 86	*Vinca minor*	*D*
pignut	20	*Conopodium majus*	*N*
pimpernel, scarlet	7 25 87	*Anagallis arvensis*	*A*
yellow	7	*Lysimachia nemorum*	*?N*
pine, Scots.	5 7 9 10 11 74	*Pinus sylvestris*	*N*
pink, Cheddar	2 61	*Dianthus gratianopolitanus*	*N*
wild	61	*Dianthus plumarius*	*A+D*
pirri-pirri bur	75	*Acaena anserinifolia*	*A*
plane, London	106	*Platanus x hybrida*	*−*
plantain, greater	34 76 81 84 85 98	*Plantago major*	*N*
ribwort	9 34 40 81 87 98	*Plantago lanceolata*	*N*

plum, wild	35 39	Prunus domestica	G
policeman's helmet or Himalayan/Indian	79 83	Impatiens glandulifera	D
pondweed, Canadian	80	Elodea canadensis	D
poplar, balsam	103	Populus trichocarpa	G
poppy, opium	26 28 35 41	Papaver somniferum	D
red or common	16 96	Papaver rhoeas	A+C
rough	25	Papaver hybridum	A
Welsh	9 44	Meconopsis cambrica	N
potentilla	47	Potentilla spp.	N
primrose, common	2 15 37 40 47 50 81 95 97	Primula vulgaris	C
evening	59 111	Oenothera stricta	D
pudding grass or pennyroyal	28 40 61 84 95	Mentha pulegium	D
quince	35	Cydonia oblonga	D
ragwort, common	81 87	Senecio jacobaea	N
Oxford	63 81 94 97	Senecio squalidus	D
rampion, harebell	41	Campanula rapunculus	D
rape, oil seed	47	Brassica napus	–
redwood	5	Sequoia sempervirens	D
reed	110	Juncus spp.	N
reedmace	76	Typha spp.	N
rhododendron	73 109	Rhododendron ponticum	D
rocket, yellow or wintercress	102	Barbarea vulgaris	N
rose, christmas	35	Helleborus niger	G
rock	10 19	Helianthemum nummularium	N
wild or dog	39	Rosa canina	N
rosemary	2 30 54	Rosmarinus officinalis	N+D
rowan	5 11 36	Sorbus aucuparia	N+D
rue	30	Ruta graveolens	G
rush, flowering	76	Butomus umbellatus	N
rye	1 33	Secala cereale	–
sage	2 29	Salvia officinalis	G
sainfoin	47 57	Onobrychis viciifolia	N+?C
samphire, rock	53	Crithmum maritimum	N
sandwort	20	Arenaria montana	N
saxifrage	10	Saxifraga spp.	N
opp.-leaf golden	10 62	Chrysosplenium oppositofolium	N
scabious, field	15 47 59	Knautia arvensis	N
sea pink or thrift	10	Armeria maritima	N
self heal	12 40	Prunella vulgaris	N
service, wild	11 37 39 47	Sorbus torminalis	N
shepherd's purse	86	Capsella bursa–pastoris	N
sloe or blackthorn	5 39 71 97	Prunus spinosa	N
sneezewort	40	Achillea ptarmica	N
snowberry	78	Symphoricarpos rivularis	D
soapwort	41 43 85 97	Saponaria officinalis	D
solomon's seal	53	Polygonatum multiflorum	N
soldier's pride or valerian	55 76	Valeriana officinalis	D
sorrel, sheeps'	13 36	Rumex acetosella	N
wood	99	Oxalis acetosella	N
sowthistle	25 86 87 88	Sonchus oleraceus	A
spearmint	61	Mentha spicata	D
spearwort, lesser	76	Ranunculus flammula	N
speedwell	40	Veronica spp.	N
common field	77	Veronica persica	D
slender	96	Veronica filiformis	A
spindle	12 39 47	Euonymus europaeus	N
spring beauty	80	Montia perfoliata	D

spruce, Sitka	107	*Picea sitchensis*	–
spurge	7	*Euphorbia spp.*	N
petty	7 87	*Euphorbia peplus*	N
spurrey, corn	18 25 34	*Spergula arvensis*	C
St John's wort	7 40 87	*Hypericum spp.*	N
star of Bethlehem	59 85	*Ornithogalum umbellatum*	?N
storksbill musk	85	*Erodium moschatum*	N
strawberry tree	10 60	*Arbutus unedo*	G
sundew	50	*Drosera rotundifolia*	N
sweet briar	87	*Rosa rubiginosa*	N
swinecress	25	*Coronopus squamatus*	A
sycamore	29 90	*Acer pseudoplatanus*	D
tamarisk	51	*Tamarix gallica*	D
tansy	41 42 81 84	*Tanacetum vulgare*	D
teazle	44	*Dipsacus pilosus*	N
Thanet cress	77 86 87	*Cardaria draba*	A
thistle, creeping	76 81 86	*Cirsium arvense*	N
spear	76 81 87	*Cirsium vulgare*	N
thorn apple	55	*Datura stramonium*	D
thrift or sea pink	10	*Armeria maritima*	N
thyme	2 29	*Thymus vulgaris*	D
wild	19 53 101	*Thymus serpyllum*	N
toadflax, ivy-leaved	60 63 76 81 83	*Cymbalaria muralis*	D
common	85	*Linaria vulgaris*	N
tormentil	20	*Potentilla erecta*	N
townhall clock	1	*Adoxa moschatellina*	N
traveller's joy or wild clematis	9 50	*Clematis vitalba*	N
trefoil, birdsfoot	1 11 15 19 40 72 103	*Lotus corniculatus*	N
tutsan	91 102	*Hypericum androsaemum*	N
twayblade spp.	54	*Listera cordata/ovata*	N
valerian or soldier's pride	55 76	*Valeriana officinalis*	D
vervain	59	*Verbena officinalis*	N
vetch	12 15 72 81 87	*Vicia sativa*	N
vine	3 35 37 42	*Vitis vinifera*	G
violet, native or dog	40 99	*Viola riviniana*	N
sweet	40	*Viola odorata*	N
Teesdale	20 78	*Viola rupestris*	N
wallflower	2 40 41 54 55 102	*Cheiranthus cheiri*	A
walnut	29	*Juglans regia*	D
waterlily, fringed	57	*Nymphoides peltata*	A
waybread or greater plantain	34	*Plantage major*	N
wayfaring tree	50	*Viburnum lantana*	N
weld	93	*Reseda luteola*	D
wheat, bread	1 16 33 56 68 85	*Triticum aestivum*	–
willow	7 9 11 97 107 110	*Salix spp.*	N
dwarf	7 9 11	*Salix herbacea*	N
willowherb	12 102	*Epilobium spp.*	N
American	96	*Epilobium adenocaulon*	A
wingnut	5 6	*Pterocarya fraxinifolia*	G
wintercress or yellow rocket	102	*Barbarea vulgaris*	N
woad	9 92 97	*Isatis tinctoria*	N
woodruff	91	*Galium odoratum*	N
yarrow	81 86	*Achillea millefolium*	N
yellow rattle	20	*Rhinanthus minor*	N
yellow wort	108	*Blackstonia perfoliata*	N
yew	5 11	*Taxus baccata*	N

SELECTED FURTHER READING

Allen, Mea: *The Tradescants*, Michael Joseph 1964
Crawford, Peter: *The Living Isles*, BBC 1985
Fitter, R.S.R.: *London's Natural History*, Collins 1945
Gerard, John: *The Herbal*: enlarged & amended: Thomas Johnson. 1663
Gilmour, John, & Walters, Max: *Wild Flowers*, Collins 1945
Godwin, Harry: *History of British Flora*, CUP 1975
Grigson, Geoffrey: *The Englishman's Flora*, Phoenix 1955
Hoskins, W.G.: *Making of the English Landscape*, Hodder & Stoughton 1955
Mabey, Richard: *Street Flowers*, Penguin 1976
Mabey, Richard: *The Common Ground*, Hutchinson 1980
Rackham, Oliver: *Trees and Woodland in the English Landscape*, Dent 1990
Rackham, Oliver: *Illustrated History of the Countryside*, Wiedenfeld & Nicholson 1994
Rhind, William: *A History of the Vegetable Kingdom*, Blackie & Son 1860
Salisbury, Edward: *Weeds and Aliens*, Collins 1961
Shoad, Marion: *The Theft of the Countryside*, Bath 1980
Tansley, A.G.: *Britain's Green Mantle*, Allen & Unwin 1949
Taylor, Christopher: *Roads and Tracks of Britain*, Dent 1979
Turrill, W.B.: *British Plant Life*, Collins 1946
Tusser, Thomas: *Good Points of Husbandry*, 1557, ed. Hartley. M., Oxford

Cappella Archive
Limited Editions

Cappella Archive provides a similar mastering service for the written word that a recording studio does for music. The typeset book file is stored in a digitized Archive and copies are printed on request as they are ordered; the Archive behaving as the printing equivalent of audio or video dubbing.